Chemical Products and Processes

Konrad Hungerbühler • Justin M. Boucher •
Cecilia Pereira • Thomas Roiss •
Martin Scheringer

Chemical Products and Processes

Foundations of Environmentally Oriented Design

Konrad Hungerbühler
ETH Zürich
Institute for Chemical and Bioengineering
Zürich, Switzerland

Justin M. Boucher
ETH Zürich
Institute for Chemical and Bioengineering
Zürich, Switzerland

Cecilia Pereira
EBP Schweiz AG
Zürich, Switzerland

Thomas Roiss
Vienna, Austria

Martin Scheringer
ETH Zürich
Institute of Biogeochemistry & Pollutant
Dynamics
Zürich, Switzerland

RECETOX
Masaryk University
Brno, Czech Republic

ISBN 978-3-030-62424-8   ISBN 978-3-030-62422-4   (eBook)
https://doi.org/10.1007/978-3-030-62422-4

Translated and fully revised from the original German language publication: Chemische Produkte und Prozesse by Konrad Hungerbühler, Johannes Ranke, Thomas Mettier. Copyright Springer-Verlag Berlin Heidelberg 1999. All rights reserved.

This Springer imprint is published by the registered company Springer Nature Switzerland AG.
The registered company address is: Gewerbestrasse 11, 6330 Cham, Switzerland

This book is dedicated to all of the research staff and students who were part of the Safety and Environmental Technology Group at ETH Zurich from 1994 to 2018.

Thank you for your curiosity and commitment to help better understand and improve the chemical products and processes of today to ensure a safer and more responsible world tomorrow.

Foreword

Twenty-five years ago, Professor Konrad Hungerbühler started a new professorship for Safety and Environmental Technology with a focus on chemical products and processes at the Swiss Federal Institute of Technology (ETH Zurich). Funding to support the research under this professorship was donated by the Swiss chemical industry in the wake of a major chemical accident in 1986 at the Schweizerhalle industrial complex in Basel, Switzerland, which necessitated more careful management of the safety and environmental effects of chemical products and processes. For this, not only thorough scientific research was needed but also an extensive education of students in chemistry, environmental science, and process engineering.

This book documents the state-of-the-art concepts in this discipline after 25 years of the research group's intensive work with numerous students, scientific co-workers, peers, and industry experts. It gives a broad overview of the basics and core elements of safe and environmentally responsible chemical products and processes, including:

- Sustainability as a balance between safety and environmental protection and societal and economic aspects
- The precautionary principle in regard to uncertainty
- Risk assessment and risk management, risk dialogue, balancing risks, and benefits
- Optimizing products and processes for the lowest overall negative impact
- Thermal process safety

This book is designed not only for safety and environmental specialists but for all those who want to or do work in the process industry and, in particular, in chemical product and process development. It especially aims to help educate current students in the fields of chemistry and environmental science as well as chemical and environmental engineering. These students should become the next generation of responsible leaders in their roles within industry, government agencies, politics, academia, and beyond. With its concrete practical examples and case studies, readers learn not only the theoretical aspects but also how to practically implement these concepts. Graduates will enter the industry with a thorough mindset of safety and environmental protection, which is what is needed to avoid chemical incidents and environmental damage, thus maintaining and increasing public confidence in the chemical and process industry. Society highly benefits from chemical products in all areas of life, but will only continue to accept them when the highest standards are applied. This book lays the foundation for protecting people and the environment by preventing damage from chemical products and processes.

I thank Professor Hungerbühler, the coauthors, and all contributors from the ETH Zurich Safety and Environmental Technology Group for this invaluable textbook, and I hope that it will help attain the goal of educating the next generations of professionals on how to design safe and environmentally responsible chemical products and processes for the benefit of society.

Group Chief Officer   Dr. Peter Schnurrenberger
Safety, Security, Health, and Environmental Protection
F. Hoffmann-La Roche AG
4002 Basel, Switzerland
January 2021

Preface

Quality of life in today's industrialized world is relying increasingly on intensifying the production and use of synthetic chemicals. Behind this development is an enormous global consumption of natural resources and energy as well as an immense production of waste and emissions. Further, many of these chemicals pose acute and chronic risks along their entire life cycle to the well-being of human health and the environment. To contribute to a sustainable future and considering this anticipated growth, a rigorous and environmentally oriented (re-)design of chemical technology is urgently needed.

Achieving such a fundamental paradigm shift requires bringing together knowledge and tools from across disciplines and teaching them to a wide range of scientists, engineers, and decision-makers. This book aims to provide readers with this transdisciplinary, comprehensive foundation of the environmentally oriented design of chemical products and processes. Divided into three parts, it first provides a normative foundation introducing the concepts of technology, risk, sustainability, and legislation and the guiding principles of integrated development. The second part then helps readers build a technical foundation, guiding them through the use of the methods of life cycle assessment, product and process risk assessment, and risk-benefit dialogue. Finally, the last part of this book provides practical guidance on implementing these methods and tools through a fully workable case study developed with industry that investigates several aspects of a real crop protection product.

Throughout the chapters, the concept of integrated development is used, and emphasis is placed on applying its guiding principles beginning in the earliest design stages, which benefit from a high degree of freedom but are often limited by the scarcity of needed data. Development is encouraged as an iterative process where critical scenarios are modeled, and variability and uncertainty are considered using both qualitative and quantitative treatments.

The most important audience for this book is perhaps simply best described as anyone interested in helping to develop or make use of safer and more responsible chemical technology—with the aim to ensure that terms such as "green" and "sustainable" are applied to comprehensive solutions, rather than just being used as catchwords. This includes professionals from research and development within the chemical industry, academia, national laboratories, civil society organizations, or wherever new chemical technology is being developed or critically reviewed. This also certainly includes students and their instructors, particularly on the master's level across chemistry, engineering, and environmental science disciplines.

Given the broad range of often complex topics that this book addresses, it is not intended to be a resource for specialists within single, specific fields. Instead, the focus is placed on presenting readers with many of the fundamental and most important principles within each field without getting lost in the discipline-specific details. For those interested in a deeper discussion and understanding of specific concepts, links to relevant literature sources are provided.

The transdisciplinary approach taken in this book is unique. It does not require readers to have extensive theoretical or technical background knowledge, but it does demand that they have the willpower to take a step back and examine the bigger picture. Compared to 20 years ago when the first edition of this book was published, the world has continued to rapidly globalize with increasing volumes of ever-more complex chemical products and processes entering the market. Safety and the protection of human health and the environment are therefore as (if not more) important to carefully consider today as they were then. I hope this book can serve as a useful resource for those interested in helping to continuously improve the responsible use of chemicals in the decades to come.

Zürich, Switzerland   Prof. Konrad Hungerbühler
January 2021

Acknowledgments

Writing this book was made possible thanks to the support from many of our colleagues across the multiple disciplines covered in this book. Specifically, we would like to thank Dr. Christoph Baumberger for providing helpful comments on the philosophical elements touched upon in the second chapter. Both Dr. Peter Schnurrenberger and Dr. Patric Bieler are acknowledged and thanked for providing us with helpful insights into the latest product and process development practices within the pharmaceutical and chemical industries. For the development of the main case study, we would like to thank both Dr. Georg Geisler and Dr. Wolfgang Stutz for their valuable feedback. The chapter on life cycle assessment greatly benefited thanks to the feedback and invaluable discussions with Prof. Dr. Stefanie Hellweg. We would also like to thank Dr. Kaspar Eigenmann for his helpful input provided during the development of the chapter on societal dialogue.

The presentation of the topics included in this book were formulated and refined over the course of many years of research and teaching at ETH Zurich. This was only possible thanks to the significant support from many former teaching assistants and research scientists within our group. We would like to give a very special thanks to these colleagues and all of our former group members who contributed to the course curriculum and to advancing research in the field.

Contents

List of Abbreviations

atm	Atmosphere
bw	Body weight
ca.	Circa, approximately
Chap.	Chapter
d	Day
eq.	Equivalents
Eqn.	Equation
Fig.	Figure
ha	Hectare
hr	Hour
L	Liter
min	Minute
p.	Page
Sect.	Section
stst	Steady state
t	Metric ton or tonne
yr	Year

Acronyms

ADI	Acceptable Daily Intake
ADME	Absorbed, Distributed, Metabolized, and Excreted
AF	Assessment Factor
AICHE	American Institute of Chemical Engineers
ALARP	As Low as Reasonably Practicable
AOEL	Acceptable Operator Exposure Level
AOX	Adsorbable Organic Halides
ASM	Acibenzolar-S-Methyl
ATP	Adenosine Triphosphate
BAF	Bioaccumulation Factor
BAT	Best Available Technology
BCF	Bioconcentration Factor
BMF	Biomagnification Factor
CAD	Chemical Agents Directive (EU)
CAS	Chemical Abstracts Service
CCA	Chemical Control Act (South Korea)
CCPS	Center for Chemical Process Safety
CED	Cumulative Energy Demand
Cefic	European Chemical Industry Council
CEPA	Canadian Environmental Protection Act
CF	Characterization Factor
CFCs	Chlorofluorocarbons
CH	Switzerland (Confoederatio Helvetica)
CHMP	Committee for Medicinal Products for Human Use
CLP	Classification, Labelling and Packaging Regulation (EU)
CMD	Carcinogens and Mutagens Directive (EU)
CMR	Carcinogenic, Mutagenic, or Toxic for Reproduction
COD	Chemical Oxygen Demand
COMAH	Control of Major Accident Hazards
CPD	Committee for the Prevention of Disasters
CSCL	Chemical Substances Control Law (Japan)
CSR	Chemical Safety Report
CTU	Comparative Toxicological Unit
DALY	Disability-Adjusted Life Years

DDT	Dichlorodiphenyltrichloroethane
DNEL	Derived No-Effect Level
DOC	Dissolved Organic Carbon
DSC	Differential Scanning Calorimetry
DSL	Domestic Substances List (Canada)
EC	European Community
ECETOC	European Centre for Ecotoxicology and Toxicology of Chemicals
ECHA	European Chemicals Agency
EDC	Endocrine-Disrupting Chemical
EE	Eco-efficiency
EEA	European Environment Agency
EFSA	European Food Safety Authority
EHS	Environment, Health, and Safety
EMA	European Medicines Agency
ESCIS	Expert Commission for Safety in the Swiss Chemical Industry
ETA	Event Tree Analysis
EU	European Union
EU-OSHA	European Agency for Safety and Health at Work
FDA	Food and Drug Administration (United States)
FMEA	Failure Modes and Effects Analysis
FOEN	Swiss Federal Office for the Environment
FTA	Fault Tree Analysis
GHS	Globally Harmonized System of Classification and Labelling of Chemicals
GMP	Good Manufacturing Practice
GWP	Global Warming Potential
HAZOP	Hazard and Operability
HRA	Human Reliability Analysis
ICCA	International Council of Chemical Associations
IEC	International Electrotechnical Commission
IED	Industrial Emissions Directive (EU)
IFCS	Intergovernmental Forum on Chemical Safety
IOMC	Inter-Organization Programme for the Sound Management of Chemicals
IPCC	Intergovernmental Panel on Climate Change
IPCP	International Panel on Chemical Pollution
IPCS	International Programme on Chemical Safety (WHO)
IPL	Independent Protection Layer
IPM	Integrated Pest Management
ISHL	Industrial Safety and Health Law (Japan)
ISO	International Organization for Standardization
K-REACH	Act on Registration and Evaluation of Chemical Substances (South Korea)
LCA	Life Cycle Assessment
LCC	Life Cycle Costing

LCI	Life Cycle Inventory
LCIA	Life Cycle Impact Assessment
LCSA	Life Cycle Sustainability Assessment
LOAEL	Lowest Observed Adverse Effect Level
LRTP	Long-Range Transport Potential
MAPP	Major Accident Prevention Policy
MCDA	Multi-Criteria Decision Analysis
MEA	Multilateral Environmental Agreement
MOS	Margin of Safety
MRL	Maximum Residue Level
MSDS	Material Safety Data Sheet
MTBF	Mean Time Between Failures
MTSR	Maximum Temperature of the Synthesis Reaction
MTT	Maximum Tolerable Temperature
NGO	Nongovernmental Organization
NOAEL	No Observed Adverse Effect Level
NOEC	No Observed Effect Concentration
NPV	Net Present Value
NREL	National Renewable Energy Laboratory (United States)
ODP	Ozone Depletion Potential
OECD	Organisation for Economic Co-operation and Development
OEL	Occupational Exposure Limit
OSH	Occupational Safety and Health
P_{ov}	Overall Persistence
PAH	Polycyclic Aromatic Hydrocarbon
PBPK	Physiologically Based Pharmacokinetic
PBT	Persistent, Bioaccumulative, and Toxic
PCB	Polychlorinated Biphenyl
PDI	Predicted Daily Intake
PDSCL	Poisonous and Deleterious Substances Control Law (Japan)
PEC	Predicted Environmental Concentration
PFAS	Per- and Polyfluoroalkyl Substances
PFD	Probability of Failure on Demand
PMN	Pre-manufacture Notice
PMT	Persistent, Mobile, and Toxic
PNEC	Predicted No-Effect Concentration
POCP	Photochemical Ozone Creation Potential
POP	Persistent Organic Pollutant
PPE	Personal Protective Equipment
PRTR	Pollutant Release and Transfer Register
QSAR	Quantitative Structure-Activity Relationship
R&D	Research and Development
RA	Risk Assessment
RAC	Risk Assessment Committee

REACH	Registration, Evaluation, Authorization, and Restriction of Chemicals (EU)
RIVM	National Institute for Public Health and the Environment (Netherlands)
RQ	Risk Quotient
RSC	Royal Society of Chemistry
RVNAS	Reference Value Non-acutely Toxic Active Substances
SAICM	Strategic Approach to International Chemicals Management
SDG	Sustainable Development Goal
SEAC	Committee for Socio-Economic Analysis
SETAC	Society of Environmental Toxicology and Chemistry
SIEF	Substance Information Exchange Forum
SIL	Safety Integrity Level
SIS	Safety Instrumented System
SLCA	Social Life Cycle Assessment
SVHC	Substance of Very High Concern
TEF	Toxic Equivalency Factor
TLV	Threshold Limit Value
TMR	Time to Maximum Rate
TOC	Total Organic Carbon
TRS	Temporal Remote State
TSCA	Toxic Substances Control Act (United States)
UBA	Umweltbundesamt (German Environment Agency)
UN	United Nations
UNEP	United Nations Environment Programme
UNFCCC	United Nations Framework Convention on Climate Change
UNGA	United Nations General Assembly
US CSB	United States Chemical Safety and Hazard Investigation Board
US EPA	United States Environmental Protection Agency
US	United States
UV-VIS	Ultraviolet-Visible
VOC	Volatile Organic Compound
vPvB	very Persistent and very Bioaccumulative
WHO	World Health Organization
WWTP	Wastewater Treatment Plant
XPS	Extruded Polystyrene

Part I

Normative Foundation

Introduction and Overview

1

1.1 The Challenge of Competition, Safety, and Protection of Human Health and the Environment for the Chemical Industry

Since the early days of industrial chemistry at the end of the eighteenth century, *chemical products and processes* have been recognized not only to provide incredible benefits but also to come with potential dangers. Initially, society focused on understanding how *human health* can be adversely affected by new chemical technology, and later this concern expanded to also include effects on the *environment*. Beginning in the 1960s, increasing chemical production volumes in the wake of the great success of industrial chemistry, an improved level of awareness of potential hazards, and concern for growing environmental problems led to a greater focus on the importance of safety and the protection of human health and the environment. However, this only gained real significance above other priorities (such as cost-effectiveness) following the introduction of comprehensive environmental and chemical legislation in the 1970s and 1980s. To achieve long-term success in the chemical industry, today it is understood that products need to be able to generate high added value and at the same time be manufactured, used, and disposed of safely and in an environmentally sound manner.

At the onset of this shift in the 1970s, it was initially seen to bring only disadvantages to industrial chemical companies in the form of costly environmental requirements and growing constraints within the product life cycle. However, more and more companies began to recognize that it also created opportunities for them to positively differentiate themselves on the market and ensure the lasting value of their products as well as to improve their reputation. Stricter emission regulations catalyzed the restructuring of entire product portfolios toward greater energy efficiency, improved use of raw materials, and minimized wastes, which was recognized to reduce costs in the long term. The trend toward safer and more environmentally compatible chemicals aimed to avoid longer-term effects such as chronic toxicity and bioaccumulation, and this was recognized to also help prevent significant future costs associated with paying compensation for damages.

An ongoing challenge for industrial chemistry today is the implementation of safety and the protection of human health and the environment given the following: (1) Global competitiveness has become an essential aspect for the survival of businesses. (2) The chemical industry has become an embedded part of society with mutual dependencies. This requires industry not only to consider meeting the needs of customers but also to be accepted by society. (3) The natural environment provides crucial resources that society depends on, and its long-term preservation should be of concern to every decision-maker. The material and energy cycles within the environment provide a framework (but also constraints) for economic activities, and their importance increases the more their limits are exceeded.

To help overcome these challenges, this book aims to present readers with basic elements and techniques that can be used to ensure safety and the protection of human health and the environment within the development of chemical products and processes. These are introduced in the context of applying *integrated development*, which is a core concept used throughout the chapters to come.

1.2   The Concept of Integrated Development

Integrated development is the approach of considering risks related to safety and the protection of human health and the environment in the development of a product and its associated production processes. This includes risks from throughout the entire life cycle of the product and process, which should be considered from the very first development steps. Integrated development makes use of inputs from multiple perspectives and areas of expertise, and it encourages iterative improvements as illustrated in Fig. 1.1.

Within integrated development, every development stage of the planned technical system needs to be designed, modeled, analyzed, and then evaluated. This requires that a minimal set of relevant data be either newly generated or collected. In particular, the modeling step should aim to include data about material and energy flows that characterize interactions the system will have with humans and the environment throughout the entire life cycle. The analysis step then aims to identify both adverse effects and benefits of the system according to different scenarios that could occur. In a final step, an evaluation is carried out of the expected benefits, adverse effects, and risks considering economic objectives, legal framework conditions, societal values, and the company's desired profile. This model-based approach allows for a comparison of each potential product or process design by creating a profile of strengths and weaknesses for each variant.

For an effective and efficient development process, the widest possible range of design variations (e.g. products, synthesis processes, application processes, recycling technologies, utilization/disposal routes) should be initially considered. Further and more refined versions of the technical system can then be created for

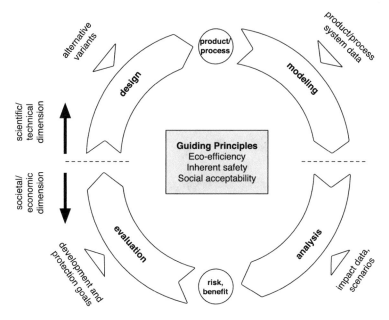

Fig. 1.1   Schematic of integrated development as an iterative process

an improved estimation of the desired and undesired effects and a more informed assessment. Within integrated development, three guiding principles are used in the modeling, analysis, and evaluation of design variants: (1) eco-efficiency, (2) inherent safety, and (3) social acceptability. These guiding principles are introduced in more detail later in Chap. 4.

1.3   Learning Goals

This book has been developed based on the lecture 'Risk Analysis of Chemical Processes and Products', which was a part of the study program for chemists, chemical engineers, and environmental scientists at ETH Zurich in Switzerland for over 20 years. It brings together a diverse set of fundamental concepts often taught across a range of standard university courses such as environmental chemistry, process safety, and general toxicology. Readers are expected to already have a solid foundation of general scientific and technical knowledge, and, due to the interdisciplinary nature of safety, health, and environmental protection (see Fig. 1.2), basic knowledge in other areas such as economics and chemical engineering is also useful to have.

Despite uncertainties, conflicting goals, and time pressure, decisions for the design of chemical products and processes need to be taken "here and now", and these decisions should be as scientifically sound as possible. This book aims to provide insight into science-based decision-making for safe chemical technology,

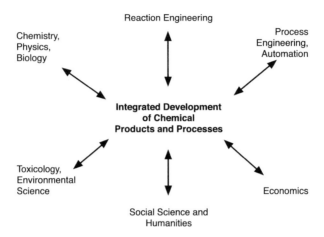

Fig. 1.2 The interdisciplinary nature of safety and the protection of human health and the environment

and it emphasizes how integrated development can help to (1) evaluate design decisions from different points of view and (2) carefully consider information early on to help ensure that the real dimensions of a problem are more readily recognized and understood. Today's chemists, chemical engineers, environmental engineers, and environmental scientists have a responsibility to lead the development of eco-efficient, inherently safe, and socially acceptable chemical processes and products. Accordingly, the main learning goals of this book are:

- To learn about the history of integrated development and recognize the achievements made and the ongoing challenges that exist
- To understand, on a mechanistic level, the basic hazards of chemical products and processes regarding safety and protection of human health and the environment
- To gain an understanding of the methods and tools used for qualitative and quantitative life cycle assessment and risk assessment of chemical products and processes
- To understand the stakeholders involved in integrated development and the approach to effective societal dialogue
- To develop and strengthen solution-oriented thinking skills and interdisciplinary teamwork

1.4   Structure and Content

To effectively apply integrated development, it is important to have both orientational knowledge about its conceptual and philosophical basis and about the context for applying it, as well as technical knowledge about the methods and tools used to analyze and manage the risks involved.

The first part of this book (Part I: Normative Foundation) includes chapters that focus on developing an interdisciplinary perspective on the field. Integrated development is discussed from an ethical, societal, and business standpoint. This part addresses questions about responsibility, motivation for iterative design, relevant legal frameworks, and fundamental business aspects for the consideration of safety and the protection of human health and the environment.

The second part (Part II: Technical Foundation) focuses on presenting a set of key technical methods and tools for applying integrated development. It includes chapters that introduce life cycle assessment, product and process risk assessment, thermal process safety, and societal dialogue. A practical set of guidelines for applying these within the integrated development of chemical products and processes is also given.

In the final part of this book (Part III: Implementation), a workable case study is presented that brings together the earlier chapters. It allows readers to apply the overarching, orientational concepts presented in Part I along with the technical methods and tools presented in Part II to analyze part of the production and use of a real crop protection product.

There are several topics relevant to the ongoing development of the chemical industry that are not specifically covered in this book. These include, for example, nanomaterials and nanotechnology; biotechnology and, in particular, the use of genetically modified organisms in chemical production; and the problem of plastic pollution. Many of the fundamental concepts presented in this book are also relevant to these topics and can be used to address them. However, these topics also require an understanding of additional, topic-specific concepts, mechanisms, and data that go beyond the scope of what can be presented in this book.

Technology, Risk, Precaution, and Sustainability

2

2.1 Natural Science and Technology as the Basis of Industrial Development

The development of modern society is strongly influenced by the progress of science and technology. Achievements over the past 150 years have brought material wealth to large fractions of the population in industrialized nations, and science and technology play a central role in reducing risks to life and avoiding scarcity within the framework of the modern economy. With today's diversity of technical products, processes, and services, however, unintended consequences such as resource depletion, the release of hazardous chemicals, and the risk of serious accidents are an ongoing and often serious threat to human and environmental health and well-being.

To really understand the importance of science and technology and also how they relate to one another, it is helpful to take a look back at their origin and historical development.

Science is the systematic study of the natural world using observations and experiments (including computer simulations). Since antiquity, science has been transformed from an intellectual and contemplative form into a strongly action-oriented approach through the inclusion of technical elements. The introduction of using experiments and the growing emancipation of humans from nature played a significant role in this process. In the European Renaissance (16th and 17th century), the idea of domination and control of nature superseded a natural science that was strongly connected with philosophy and theology. Francis Bacon (1561–1626), in his book *Novum Organum*, laid the foundation of the modern, empirically based, scientific method. Since then, a large part of knowledge-oriented science has been systematically put into the service of application-oriented science. The ability of humans to emancipate themselves from the natural environment and to change it has led, to some extent, to an antagonistic relationship between humans and nature (Hösle, 1994; Primas, 1992; Sachverständigenrat für Umweltfragen, 1994).

© The Author(s), under exclusive license to Springer Nature Switzerland AG 2021
K. Hungerbühler et al., *Chemical Products and Processes*,
https://doi.org/10.1007/978-3-030-62422-4_2

The present state of *technology* would be inconceivable without the development of science. Modern technology is the application of scientific knowledge for practical purposes to support human life. From this perspective, nature is available to satisfy human needs through the use of tools and processes.

With the ongoing development of technology, the focus to meet human needs (or desires) has become even greater, and as a consequence, also the complexity of new technology and its implementation and widespread use has increased. Some examples of technologies used today for the production of chemical products are:

- Chemical technology used for product manufacture through reactive conversion, including the control and handling of toxic, flammable, and corrosive chemicals
- Process technology used for the industrial production of substances by technical methods such as fractionated distillation or real-time monitoring and advanced treatment of wastewater
- Safety and environmental technology used for proactive risk reduction, such as inherently safe process conditions (low pressure and temperature, reactions facilitated by high-performance catalysts, etc.)

2.2   Impacts of Technology

2.2.1   Ambivalence of Technical Progress

In addition to improving the quality of life for humans, the technologization of the world is also associated with problems. For example, it:

- Allows for a concentration of power with subsequent global economic, ecological, and social imbalances
- Stimulates demands and wishes without consideration for compromise and alignment
- Creates new dangers and risks that often emerge from unintended and unforeseen side effects
- Causes strong impacts on the natural environment through the extraction of resources and the release of waste
- Leads to a lack of orientation within, and understanding of, the complexity of modern societies

Technical progress is associated with uncertainties. The increase of prosperity that springs from technical progress is connected with new *hazards*, *dangers*, and *damages*. Within the context of chemical processes and products, a *hazard* is an inherent property of a substance or situation with the potential to cause harm. A *damage* is the impairment of the inherent qualities of a *protected good*, which are valued subjects, resources, or conditions such as humans, the environment, property, etc. A *danger* is defined as a situation or circumstance in which damage may occur.

Unintended impacts can occur over the entire life cycle of a product (see Fig. 2.1). The result of industrial chemical production, for example, is not just a chemical product with its desired effect or function but also any associated side effects. While it is already difficult enough to find and explain the desired effects of a product, a clear understanding of its side effects is often many times more complex. Here, the linearity of cause-and-effect thinking is broken up into a network of possible side effects not just in terms of environmental or health impacts but also including changes of consumers' preferences, more visible roles of environmental nongovernmental organizations (NGOs), lawsuits, etc. (Beck, 1992; Grunwald, 2002).

The dilemma of technological as well as societal progress is that it can endanger the safety and prosperity of present and future generations by accepting new technical and ecological risks in the pursuit of material well-being and safety. Compared to the risks humans faced hundreds of years ago, the modern risks faced today differ not only in terms of time and space but also in psychological and social terms. For example, today's risks:

- Are global and can cross societal boundaries and even relate to the biosphere as a whole
- Threaten present and, to a certain extent, also future generations
- Have a complexity that often prevents detection, and identification of the source, of a problem
- Often elude direct perception by the human senses

Risk has become a key factor in the context of the uncertainty of technical progress. To better understand, the term itself should be more formally introduced.

2.2.2 The Concept of Risk

Risk is used in everyday language in various contexts. For example, the words "danger," "luck," and "thrill" can all be associated with the term risk and have meanings with both positive and negative attributes. In economics, risk is understood as the variance of an average value, i.e., it has both positive and negative effects. Risk as a technical term as it is used here, however, describes only *undesirable* possible future events or outcomes.[1]

Accordingly, risk is understood in the context of chemical products and processes as the potential (probability) for harm (impact) to protected goods by specific,

---

[1] In the different fields covered in this book, sometimes different conventions are used for the names of important concepts and for the symbols of variables. This includes, for example, the terms "impact" and "consequences" for an (adverse) outcome of an event and the symbols p and P for the probability of an event. Because these conventions are well established in the respective fields, they are used in different chapters of this book in parallel and without harmonization.

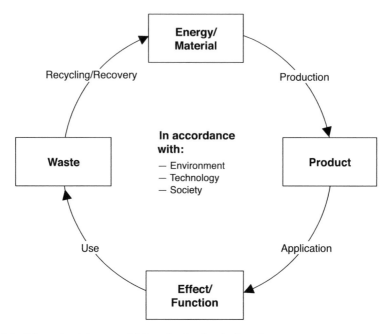

Fig. 2.1 Schematic of the general life cycle of a chemical product

hazardous conditions or circumstances. The risk (R) of a possible damage event can, therefore, be expressed in the dimensions of its probability of occurrence (p) and potential impact (I):

$$R = f(p, I) \tag{2.1}$$

When describing the risk as the expected value of the damage according to the basics of a probability calculation, the following product formula applies:

$$R = p \times I \tag{2.2}$$

which turns into a sum when different, mutually independent potential damage events j are considered:

$$R = \sum_{j} p_j \times I_j \tag{2.3}$$

Such a representation of risks requires (i) the definition of a system that allows for consideration of possible damage; (ii) damage scenarios that can be quantitatively assessed and, if applicable, aggregated; and (iii) the estimation of probabilities of occurrence for these scenarios using both empirical data (based on the relative frequency of certain events in the past) and prospective probabilistic comparisons based on logical considerations and the stochastic laws.

Defining the system boundaries as well as the assessment of specific scenarios is normative, i.e., not determined by empirical findings and data, and depends heavily on who defines the damage. From a different perspective, damage to a protected good can be seen as a benefit. Moreover, the average expected value of the loss does not allow for any objective comparison of various damage categories such as personal injury, material damage, or environmental damage. Comparing damage categories requires weighting each of them based on a system of values, which can vary significantly in a pluralistic society.

Once the risk (or risks) have been determined (also quantitatively), it is important to also consider the various ways in which a risk can be perceived. The same risk (in terms of its probability and impact) can be perceived very differently between individuals or between cultures. This perception can be separated into different layers based on the extent of technical rationality applied:

1. *Scientific/technical layer of risk:* Risk calculated using only scientific and technical information. This is the most technically rational perception of risk.
2. *Psychological layer of risk:* Risk expressed using an individual's personal understanding of a risk (including subjective biases and preferences). At this level, technical rationality is no longer sufficient to cover all aspects that may be perceived as a risk by a person.
3. *Sociological layer of risk:* Risk considered in the context of a societal understanding of risks and how they are perceived and valued by different societal groups. At this level, technical information and rationality may play a considerably less important role.

Accordingly, a technical risk assessment is often in contrast to the subjective risk perception of a person or a societal group, and there is usually no direct correlation between them. This is in part because individuals include additional dimensions in their intuitive risk perception, for example, the possibility of influencing the risk themselves or a personal connection to what is being protected. This societal context of risks should therefore not be left out. When industrial risks are perceived and allocated to a specific industry, the acceptance of these risks by society dwindles. Risks with all of these facets have become important aspects within our society (Beck, 1992). The societal perception of risks and the importance of communicating risks with society are further discussed in Chap. 9.

2.2.3 Limits to the Concept of Risk and the Precautionary Principle

A risk as it is defined in the previous section can only be determined when both the type of event (and its impact) and the probability of this event are known. In many cases, however, the probabilities of events with adverse impacts cannot be determined, or even the relevant events themselves are not yet known, for example, when a technology is newly introduced and there is no body of experience describing its typical failures and their consequences. But even well-known technologies may lead to events that nobody could anticipate and that are not known before they happen.

When events and their impacts are known, but the associated probabilities are not, this is called a situation under *uncertainty* (Wynne, 1992; Hansson, 2009).[2] For situations where neither the relevant events nor their probabilities are known, the terms "indeterminacy" (Wynne, 1992), "deep uncertainty," "great uncertainty," or "ignorance" (Hansson and Hirsch Hadorn, 2016) are used.

Importantly, many cases where one has to deal with the unwanted impacts of a technology are situations under uncertainty or even under deep uncertainty. One example is the depletion of stratospheric ozone by chlorofluorocarbons (CFCs), which was a later discovery and not at all anticipated when CFCs were introduced. Decision-making under (deep) uncertainty requires different techniques than decision-making under risk, where normally the methods of *probabilistic risk assessment (PRA)* are used. Hansson and Hirsch Hadorn (2016) present the approach of the *argumentative turn* as a method that can take into account poorly characterized outcomes, their implications, and their relevance for decision-making. It is important to acknowledge the (deep) uncertainty of many decision-making situations in the area of technology assessment and not to force the blueprint of probabilistic risk assessment on them. Probabilistic risk assessment has very high demands on the information that characterizes the events to be considered and is often not the method of choice.[3]

A concept that is often invoked in situations with a fundamental lack of knowledge and (deep) uncertainty is the *precautionary principle*. The precautionary principle is stated prominently in Principle 15 of the Rio Declaration of 1992: "In order to protect the environment, the precautionary approach shall be widely applied by States according to their capabilities. Where there are threats of serious or irreversible damage, lack of full scientific certainty shall not be used as a reason for postponing cost-effective measures to prevent environmental degradation" (UNGA, 1992).

The role of the precautionary principle in the context of poorly known environmental impacts of technology has been analyzed in two landmark reports on *Late Lessons from Early Warnings* published by the European Environment Agency (2002, 2013). In the two reports, more than ten chemical-related case studies are analyzed, including asbestos, polychlorinated biphenyls, lead in gasoline, perchloroethylene in plastic water pipes, mercury in fish (the Minamata case), and

---

[2]As it was discussed for the term "risk" in the previous section, also "uncertainty" is used here in a specific technical sense, but it can also be understood in a more general way (Hansson and Hirsch Hadorn, 2016).

[3]Hansson (2009) introduces the term "tuxedo fallacy" (after the gambler in a casino wearing a tuxedo and calculating probabilities) for cases where a probabilistic risk assessment is attempted although the situation is one under (deep) uncertainty.

DDT. In all of these cases, early indications of potential problems were available, but action was taken only slowly and/or to an insufficient extent. In other words, precautionary action often would have been appropriate and would have prevented serious damage. The cases show that the cost of inaction can be very high, including direct financial costs for companies. On the other hand, only a very small number of false positives were identified, i.e., cases where the early warnings did not indicate major problems to come. A list of 12 "late lessons" is presented in both reports, and already the first three of them (quoted below) highlight how important it is to acknowledge the presence of incomplete knowledge, deep uncertainty, and unexpected outcomes (p. 168–169 in European Environment Agency (2002)):

• "acknowledge and respond to ignorance, as well as uncertainty and risk, in technology appraisal and public policymaking,
• provide adequate long-term environmental and health monitoring and research into early warnings,
• identify and work to reduce "blind spots" and gaps in scientific knowledge."

Finally, there is an unresolved issue connected with the precautionary principle. On the one hand, the precautionary principle provides a convincing logical scheme that accommodates uncertainty and ignorance and supports (regulatory) action even in cases where the available evidence is not (yet) fully conclusive. It does so by emphasizing the possibility of strong adverse impacts that may become manifest and, once they have materialized, would cause (extremely) high costs.

On the other hand, there is a pervasive problem caused by the lack of conclusive evidence: the point that absence of evidence is often taken for evidence of absence. This is a logical fallacy, but this fallacy seems to be highly tempting to some scientists and decision-makers. Chemical hazard and risk assessments with limited data are often concluded with a negative outcome ("no hazard") although the insufficient data basis would require additional testing. This problem could be overcome if chemicals with insufficient data had to be labeled as "insufficiently tested" and if this label implied disadvantages for the marketing of such chemicals, as suggested by Hansson and Rudén (2006).

2.3   The Concept of Sustainability

2.3.1   Sustainable Development and Its Implementation

The concept of sustainable development was defined by the UN Commission on Environment and Development in the report *Our Common Future* (also known as the Brundtland Report) published in 1987 (UN, 1987):

Sustainable development is development that meets the needs of the present without compromising the ability of future generations to meet their own needs.

The concept of sustainability invokes the continuity of humanity as a central norm, focusing on the future satisfaction of human needs. The definition is clearly anthropocentric, with nature being significant for sustainability as the basis for the satisfaction of human needs—especially the basic needs for food, water, shelter, and hygiene. The strong emphasis on the needs of future generations is in line with the thinking of Jonas (1984), who maintains in his book, *The Imperative of Responsibility*, that humans should act—under an ethical imperative—in a way that is "compatible with the permanence of genuine human life" (Jonas, 1984).

As another key element, in recognition of the Earth as a whole system, the concept of sustainability aims to combine the development of economy and society with a recognition of the limits and constraints given by a finite environment. This concept aims to achieve a new type of progress that simultaneously satisfies economic, social, and ecological criteria.

Sustainability describes a set of requirements for the development of society's use of nature, and in this way, it becomes a guide for technical progress as well. The conditions and methods needed to achieve sustainable development are still heavily discussed even today, more than 30 years after the Brundtland Report.

The objectives of sustainability can be described by the following three basic aims:

1. *Economic:* Efficient use of resources during the creation of value for improving the overall quality of human life[4]
2. *Social:* Orientation toward humanity as a guiding factor (e.g., for values such as equality of opportunities), promotion of social cohesion and cultural identity, as well as the development of democratic institutions
3. *Environmental:* Acknowledgment of the limited capacity of the Earth's ecosystems, especially of soil, water, air, and the biosphere, in terms of their use as a source of resources and as a sink for waste and pollutants

The concept of sustainability or, more precisely, sustainable development was internationally accepted as an overarching political aim when the *Agenda 21* was signed at the UN Conference on Environment and Development (the "Earth Summit") in Rio de Janeiro in 1992 (UN, 1992). The Agenda 21 specifies in four main parts the social and economic dimensions of sustainable development and the conservation and management of resources; it addresses the roles of various societal groups in the implementation of the Agenda, such as children and youth; women; NGOs; local authorities; business, industry, and workers; farmers; and indigenous peoples, and it further outlines how sustainable development can be put into practice through education, technology transfer, research, etc.

---

[4]The current economic system is increasingly challenged by analyses that claim that its underlying mechanism of continuous growth makes it incompatible with the social and environmental aims of the sustainability concept (Daly, 2008; Seidl and Zahrnt, 2010).

There were several international follow-up meetings to the 1992 Earth Summit, including the Johannesburg Earth Summit in 2002 and the Rio +20 Summit in Rio de Janeiro in 2012, where the implementation of Agenda 21 was evaluated.

In 2015, the concept of sustainable development was given a new structure through the definition of 17 Sustainable Development Goals (SDGs) under the *Agenda 2030* (UNGA, 2015). The Agenda 2030 is in full agreement with the Agenda 21 and is intended to reaffirm the goals of the Agenda 21. The 17 SDGs are interdependent and capture the initial three dimensions of sustainable development. They provide a more highly resolved approach to the challenges of sustainable development, with each SDG being further resolved into several targets and one to three indicators per target (169 targets and 232 indicators in total).

None of the SDGs is explicitly related to impacts from chemical processes and products, but several of them have a close connection to the production and use of chemicals:

- Target 3.9 under SDG 3, ensure healthy lives and promote well-being for all at all ages, reads, "By 2030, substantially reduce the number of deaths and illnesses from hazardous chemicals and air, water, and soil pollution and contamination."
- Target 12.4 under SDG 12, ensure sustainable consumption and production patterns, says, "By 2020, achieve the environmentally sound management of chemicals and all wastes throughout their life cycle, in accordance with agreed international frameworks, and significantly reduce their release to air, water and soil in order to minimize their adverse impacts on human health and the environment."
- Target 6.3 under SDG 6, ensure availability and sustainable management of water and sanitation for all, states, "By 2030, improve water quality by reducing pollution, eliminating dumping and minimizing release of hazardous chemicals and materials, halving the proportion of untreated wastewater and substantially increasing recycling and safe reuse globally."

The relationships between the production and use of chemicals and the SDGs are described in detail in the *Global Chemicals Outlook II* published in 2019 (UNEP, 2019).

For future-oriented decisions under the SDGs, it is important to keep the many different functions of the natural environment clearly in mind. These functions include, among others (de Groot, 1992):

- Regulatory functions (regulating the climate, water and groundwater balance, soil fertility, material cycles, etc.)
- Support functions (for housing, agriculture, recreation, etc.)
- Production functions (food, genetic resources, raw materials)
- Cultural functions (such as aesthetic, spiritual/religious, historical and inspirational functions)

In the discussions about the requirements for sustainable development, a long-term controversy has been on how various resources (both natural and man-made capital) can be replaced by one another and whether or not they are substitutable. *Strong sustainability* aims to preserve the stock of environmental capital, whereas *weak sustainability* assumes that virtually all resources (including natural ones) can be substituted by other resources (Ang and Van Passel, 2012). Consequently, a decrease of a specific resource is considered justifiable by the concept of weak sustainability if functionally equivalent substitution possibilities exist and the *total* capital stock does not decrease over time. The respective definitions of environmental capital and total capital are often disputed. For a critical discussion of the sustainability concept, see Beckerman (1994), who maintains that strong sustainability is not feasible because it is too strict, whereas weak sustainability is nothing else than welfare maximization as it is known in traditional economics.

It is plausible that some resources are substitutable for individual functions (e.g., fuels), but not in regard to all functions, some of which are not even known yet. Pearce, therefore, introduced the concept of critical natural resources (Pearce et al., 1990), which are defined as resources considered so essential that they should not be allowed to diminish even though substitutes may be acceptable for other natural resources.

In the 1990s, the discussion of how sustainable development can be reached strongly focused on this question of stocks of "capital" and how they may best be used or preserved. The idea was to find guidelines that would enable societies to permanently preserve the functions of the environment as a source of resources, as the absorption medium for emissions, and as a basis for livelihood. Accordingly, early criteria for sustainable development focused on the management of stocks and flows of different types of resources (Enquete-Kommission, 1993; Interdepartementaler Ausschuss Rio, 1995; Industrie- und Handelskammer Nürnberg für Mittelfranken, 2002):

- *Criterion for the stability of renewable resources:* Consumption rate < regeneration rate. Supplies from a renewable resource may be used only at the rate at which it is created during the same time. The stock of renewable resources may not decrease.
- *Criterion for the stability of nonrenewable resources:* Consumption rate < compensation rate. Supplies from a nonrenewable resource may be used at a rate at which it can be substituted by equivalent renewable resources. The substitute must cover all essential functions of the resource.
- *Criterion for the burden from emissions:* Emission levels < exposure threshold. The emission load must not exceed the load threshold of the overall system including long-term and combined effects as well as accumulation and degradation processes.
- *Criterion for stability in ecosystems:* Rate of change < rate of adaptation. Changes in the living conditions of major ecosystems should not happen more quickly than their ability to react to such changes.

In current schemes for implementing sustainable development, such as the Swiss or German guidelines (Steinemann et al., 2019; Die Bundesregierung, 2018), this type of criteria is no longer used. The guidelines now refer directly and extensively to the Agenda 2030 and the SDGs and define country-specific "action areas." Often, defined goals include increasing resource efficiency as well as reducing material and energy footprints of production and consumption, internalization of external costs (see next subsection), increasing generation of energy from renewable sources, and the inclusion of various aspects of sustainable development in school curricula and university study programs, as well as social security and gender equality.

In an official evaluation of the current Swiss Sustainable Development Guidelines published in 2019 (Steinemann et al., 2019), it is stated that the actions indicated in the action areas have not been systematically and transparently selected, that how the actions may or may not be carried out is not governed by the Guidelines, and that there is no controlling that would indicate the actual progress (or lack thereof) toward the goals. This shows that sustainable development still is a fundamental challenge and that a real transformation of the resource- and energy-intensive socioeconomic systems of modern industrialized countries is not yet underway.

2.3.2   Sustainability and the Economy

Historically and only until fairly recently, the environment (and to some extent human health) had been on the margins of economics. Free goods such as air, water, and biodiversity were not valued but rather free of charge with nature assumed to provide practically any amount needed. They were simply not a subject of economics since they were not understood as being limited resources. Following this line of thinking, nature is expected to always deliver the needed resources and absorb the resulting waste, even for a growing human population with increasing demands. No thought was given to the carrying capacity of the planet. In the early 1970s, the limitations of natural resources and the absorptive capacity of ecosystems began to be discussed more widely as factors limiting development. This discussion started a systematic examination of economics and environmental problems (e.g., Frey (1972) and Meadows et al. (1972)). Today, natural resources such as raw materials, clean water, and clean air are undoubtedly perceived as scarce resources, even though in many places there is no functioning market governing their extraction and use.

In the 1970s, national economic markets began to show that private companies were creating additional societal costs by burdening the environment. These were not taken into account in the companies' own balance sheets because they were being carried by the general public. These are referred to as *external costs*, and, more exactly, they are external effects that occur when someone obtains a benefit at the expense of other individuals or the greater society without paying. Even today, health and environmental costs are often only partially included in a company's cost calculations. If external costs for the use of environmental resources during the production of a good are not included, the price of this good will be too low from an economic perspective. The resulting decisions of market participants then lead to misallocation of resources and ultimately to a loss in the overall economy.

The objective of a more "environmentally conscious" economy is to ensure that external costs, and therefore also related (potential) damage, are fully taken into account when pricing raw materials and products. This is one of the main goals of the Swiss Sustainable Development Guidelines (Steinemann et al., 2019).

To consider these external costs within the economy, they need to first be identified, quantified, and recorded in monetary terms (monetization) in accordance with the current state of science and technology as well as with political protection goals. In the case of *reversible* environmental damage, monetization of it can, for example, be achieved through compensation costs. For *irreversible* damage, other approaches have to be used such as an empirically determined willingness to pay for avoiding damage. The total external costs of using a good are calculated from the monetized environmental (or human health) costs and are based on the fraction that an individual polluter is responsible for.

The billing of external costs should follow the *polluter pays* principle. This is done through a political process where the calculated external costs are charged to the polluting company or entity according to the fraction that this entity is responsible for. These costs then appear in their cost calculations. This can be achieved with the help of market-oriented instruments such as:

- *Tradable emission certificates:* These certificates are marketable permissions, e.g., for certain pollutant amounts. Requirements for a meaningful certificate are (i) high geographical distribution of emissions and (ii) no resulting local toxicity. For example, they can be applied to atmospheric pollutants such as CO_2, NO_x, SO_2, or volatile organic compounds (VOCs).
- *Taxes:* Taxes could be placed on raw materials, energy, products, waste, and emissions according to their input or output quantities. As an example, in 2017, Sweden introduced a tax on various types of electronic equipment depending on the amount of brominated flame retardants present in the devices (Andersson, 2017). The resulting revenues can either be earmarked directly for the avoidance and compensation of human health and environmental damage or be reimbursed for responsible behaviors (bonus-malus system).

The funds raised through such instruments are not always necessarily used to prevent or compensate for damages. There are approaches for creating a tax system in which the increased taxation of health and environmental burdens is accompanied by a simultaneous reduction of income taxes or value-added taxes so that overall government revenues are not increased.

In an economic structure that includes health and environmental costs, the concept of health and environmental protection should not be seen as a costly fulfillment of legal requirements. Rather, it should be recognized as providing the possibility to optimize economic benefits through the efficient use of scarce resources and the protection of the public. Producers and consumers receive incentives for innovations and behavioral changes that lead to more intelligent and more sustainable use of these limited resources as well as careful use of hazardous substances. There should be a clear economic benefit to considering effects on humans and the environment. However, even after many years of political discussions, adequate prices of resources consumed and emissions generated have still not yet been introduced.

2.3.3 The Implementation of Sustainability

The concrete implementation of the concept of sustainability is seriously hampered by the complexity of economic, ecological, and societal problems. This difficulty was identified very early in the discussions about sustainable development (Sachverständigenrat für Umweltfragen, 1994), and its urgency and importance have further increased since then.

For the scope considered within this book, the focus is on approaches for the concrete implementation of sustainable practices on two different levels: (i) the technical level and (ii) the societal level.

On the technical level, the chemical industry is challenged to develop processes and products with sustainability in mind. General guidelines to achieve this can be through the implementation of high productivity, low consumption of renewable resources and energy, avoidance of emissions and waste, high levels of occupational health and safety, and minimization of accident risks and other risks. The same guidelines can be applied to the development of sustainable product life cycles.

The guidelines mentioned above are also reflected by the 12 principles of *Green Chemistry* (Anastas and Warner, 1998). Most of these principles refer to improved material and energy efficiency (principles 2, 6, 7, and 8: atom economy, design for energy efficiency, use of renewable feedstocks, reduce derivatives) and increased process safety (principle 12: inherently safer chemistry for accident prevention). Only two principles directly refer to the development of "greener" chemical products (principles 4 and 10: designing safer (i.e., less toxic) chemicals, design for degradation).

A large body of research has been published under the principle of Green Chemistry, much of it in the journal *Green Chemistry* (Royal Society of Chemistry, 2020). Several governments as well as the OECD (2019) actively support the development of Green Chemistry or Sustainable Chemistry, including the United States (US EPA, 2020) and Germany (UBA, 2020; ISC3, 2020).

However, it has to be pointed out that in practical terms, most of the research and development in the area of Green Chemistry has been put into improving chemical reactions and process conditions. Much less progress has been made in the development of truly "green" chemical products (Clark, 2006; Strempel et al., 2012; Kümmerer et al., 2020). Good summaries of the serious challenges still associated with the concept of Green Chemistry are provided by Collins (Collins, 2001; Jumman, 2013).

Even the growing acceptance and implementation of the principles of Green Chemistry, Sustainable Chemistry, and related concepts cannot resolve the impacts that will be associated with the strong projected growth of the chemical industry over the coming decades (UNEP, 2019). This problem goes beyond what a single chemical company, which is required to be economically successful in a globalized market, can resolve internally.

On the societal level, the frameworks and regulations that are necessary for the implementation of the concept of sustainability need to be strengthened. Specifically for chemicals, the concept of essential vs. nonessential uses (Cousins et al., 2019) makes it possible to evaluate which chemical products have higher priority and relevance than others. By this approach, also the problem of unchecked growth of the production of chemicals can be addressed—chemical production should be focused on chemicals for which there are essential uses.

Of general importance is the development of a sound knowledge base that facilitates the quantification of indicators that reliably measure the progress (or lack thereof) toward Sustainable Development Goals (Steinemann et al., 2019). The ability to make informed political decisions and to achieve a sound understanding of the progress made depends on having knowledge about the long-term consequences of different forms of production and consumption.

2.4   Chemical Risks and Protection Areas

This book advocates for working toward a more sustainable chemical industry by applying integrated development with support from a broad set of tools and methods. Fundamental to these is understanding that undesired impacts of chemical technology are directly (or indirectly) connected with the release of material or energy. For chemical products, this includes releases during the entire life cycle across the use, disposal, and recycling or reuse of the product. For chemical processes, this includes releases during periods of normal manufacturing as well as in the event of an accident. Impacts occur via a pathway starting with the initial release (known as the emission), followed by propagation processes, and exposure, and resulting in damage. This is depicted in Fig. 2.2, and relevant definitions are included in Table 2.1.

The impacts can be subdivided within an exposure-duration curve into the protection areas of safety, health protection, and environmental protection as shown in Fig. 2.3 and introduced by Lemkowitz (1992).

The area of *safety* concerns event-driven risks that are characterized by short exposure periods and high but mostly localized exposures. They are highly time-specific in terms of location and event. Most often, such events are considered accidents and include fires, explosions, and the discharge of toxic substances with their corresponding acute effects. This protection area is investigated and managed using the method of *process risk assessment*, which is introduced in Chaps. 7 and 8.

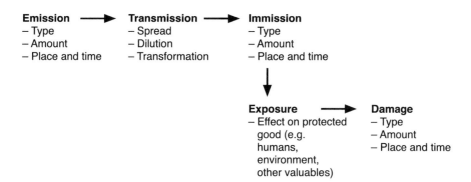

Fig. 2.2 Impact pathway from an emission to resulting damage

Table 2.1 Definition of key terms related to the impact pathway from the release of material or energy from a chemical process or product

Term	Definition
Emission	Release into the environment of non-environmental substances, energy, or organisms that act as chemical, physical, or biological stressors
Transmission	Propagation, dissipation, and conversion of an agent after emission to the environment
Immission	An agent after emission and transmission that has reached the location of the exposed protected good
Exposure	Type, magnitude, and duration of an emission at the location of a protected good
Damage	Impairment of inherent qualities of a protected good (e.g., humans, environment, property, etc.)

The area of *health protection* concerns adverse impacts on workers, consumers, and ecosystems. Risks within this area are often characterized by longer exposure periods and consequently by chronic effects. The effects are usually local but less time-specific, and the range of possible impacts is broad and may involve exposure of workers during production, side effects of consumer products, or accumulation of chemicals in environmental compartments and food chains. This protection area is investigated by means of *product risk assessment*, which is introduced in Chap. 6.

The area of *environmental protection* primarily concerns the cumulative impact of several smaller, incremental impacts distributed across space and time. These are associated with low and often unknown exposure thresholds as well as long exposure periods. Examples can include adverse effects related to product-specific wastewater emissions, exhaust air, generated waste, etc. from point sources as well as from diffuse sources. Such effects can be considered by *life cycle assessment* (LCA), which is introduced in Chap. 5. Across all three of these protection areas, the technique of *risk-benefit dialogue* is applied to ensure that risks of a chemical product or process are communicated to, and accepted by, society. This is introduced in Chap. 9.

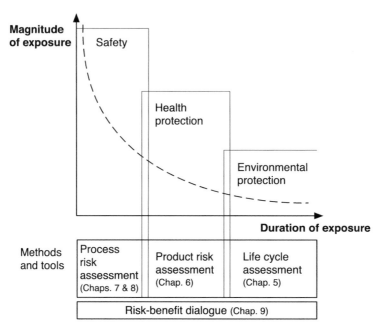

Fig. 2.3 The protection areas of safety, health protection, and environmental protection related to chemical products and processes and their classification within the magnitude-duration curve of exposure as well as the corresponding assessment methods for each. The magnitude-duration curve reflects the decrease in exposure magnitude with increasing duration. For example, this ranges from an accident within a production process (high magnitude, short duration) to a product's acute and chronic toxicity to workers (medium magnitude and duration) and finally to very long-term impacts (low magnitude, long duration)

As shown in Fig. 2.3, there are also some overlaps at the interfaces between these protection areas as well as between the methods used to address them. Some important aspects of these overlaps are introduced and discussed in Part II of this book, specifically in Chaps. 5–8.

In applying integrated development during the creation of a chemical product or process, the entire spectrum of the magnitude-duration curve should be considered, and the methods introduced in Part II are focused on this task. However, before discussing these, the important role of legislation (Chap. 3) and the embedding of integrated development into business strategies (Chap. 4) will be introduced.

References

Anastas P, Warner J (1998) Green Chemistry: Theory and Practice. Oxford University Press

Andersson AS (2017) Chemical tax can be very effective and more countries should try it. URL https://chemsec.org/chemical-tax-can-be-very-effective-and-more-countries-should-try-it/

Ang F, Van Passel S (2012) Beyond the Environmentalist's Paradox and the Debate on Weak versus Strong Sustainability. BioScience 62(3):251–259, https://doi.org/10.1525/bio.2012.62.3.6, URL https://academic.oup.com/bioscience/article-lookup/doi/10.1525/bio.2012.62.3.6

Beck U (1992) Risk Society - Towards a New Modernity, 1st edn. Sage Publishing, URL https:// uk.sagepub.com/en-gb/eur/risk-society/book203184

Beckerman W (1994) 'Sustainable Development': Is it a Useful Concept? Environmental Values 3(3):191–209, URL https://www.jstor.org/stable/30301447?seq=1

Clark JH (2006) Green chemistry: today (and tomorrow). Green Chemistry 8(1):17–21, https://doi. org/10.1039/B516637N, URL http://xlink.rsc.org/?DOI=B516637N

Collins T (2001) Toward Sustainable Chemistry. Science 291(5501):48–49, https://doi.org/ 10.1126/science.291.5501.48, URL http://www.sciencemag.org/cgi/doi/10.1126/science.291. 5501.48

Cousins IT, Goldenman G, Herzke D, Lohmann R, Miller M, Ng CA, Patton S, Scheringer M, Trier X, Vierke L, Wang Z, DeWitt JC (2019) The concept of essential use for determining when uses of PFASs can be phased out. Environmental Science: Processes & Impacts 21(11):1803–1815, https://doi.org/10.1039/C9EM00163H, URL http://xlink.rsc.org/?DOI=C9EM00163H

Daly HE (2008) A Steady-State Economy. Tech. rep., Sustainable Development Commission, URL http://www.sd-commission.org.uk/publications.php@id=775.html

Die Bundesregierung (2018) Deutsche Nachhaltigkeitsstrategie - Aktualisierung 2018. Tech. rep., URL https://www.bundesregierung.de/breg-de/service/publikationen/deutsche-nachhaltigkeitsstrategie-aktualisierung-2018-1559086

Enquete-Kommission (1993) "Schutz des Menschen und der Umwelt" des 12. Deutschen Bundestages; Verantwortung für die Zukunft: Wege zum nachhaltigen Umgang mit Stoff- und Materialströmen. Economia Verlag

European Environment Agency (2002) Late lessons from early warnings: the precautionary principle 1896–2000. Tech. rep., URL https://www.eea.europa.eu/publications/environmental_ issue_report_2001_22

European Environment Agency (2013) Late lessons from early warnings: science, precaution, innovation. Tech. rep., URL https://www.eea.europa.eu/publications/late-lessons-2

Frey BS (1972) Umweltökonomie. Vandenhoeck & Ruprecht

de Groot RS (1992) Functions of nature: evaluation of nature in environmental planning, management and decision making. Wolters-Noordhoff BV, Groningen

Grunwald A (2002) Technikfolgenabschätzung - eine Einführung, sigma edn. ITAS, Berlin, URL https://www.itas.kit.edu/pub/m/2002/grun02a_inhalt.htm

Hansson SO (2009) From the casino to the jungle. Synthese 168(3):423–432, https://doi.org/10. 1007/s11229-008-9444-1, URL http://link.springer.com/10.1007/s11229-008-9444-1

Hansson SO, Hirsch Hadorn G (2016) Introducing the Argumentative Turn in Policy Analysis. In: The Argumentative Turn in Policy Analysis, Springer, pp 11–35, https://doi.org/10.1007/978-3-319-30549-3_2, URL http://link.springer.com/10.1007/978-3-319-30549-3_2

Hansson SO, Rudén C (2006) Priority Setting in the REACH System. Toxicological Sciences 90(2):304–308, DOI 10.1093/toxsci/kfj071, URL http://academic.oup.com/toxsci/article/90/2/ 304/1658376/Priority-Setting-in-the-REACH-System

Hösle V (1994) Philosophie der ökologischen Krise, 2nd edn. C.H. Beck

Industrie- und Handelskammer Nürnberg für Mittelfranken (2002) Lexikon der Nachhaltigkeit. URL https://www.ihk-nuernberg.de/de/wir-ueber-uns/ehrbarer-kaufmann-csr/lexikon-der-nachhaltigkeit/?sid=qpdi1j853ncfhfsiivtompjuu6

Interdepartementaler Ausschuss Rio (1995) Elemente für ein Konzept der nachhaltigen Entwicklung - Diskussionsgrundlage für die Operationalisierung

ISC3 (2020) International Sustainable Chemistry Collaborative Centre (ISC3). URL https://www. isc3.org/en/home.html

Jonas H (1984) The Imperative of Responsibility. University of Chicago Press, URL https://www. press.uchicago.edu/ucp/books/book/chicago/I/bo5953283.html

Jumman A (2013) Terry Collins - Green Chemist. URL http://advancinggreenchemistry.org/terry-collins-green-chemistry-leader/

Kümmerer K, Clark JH, Zuin VG (2020) Rethinking chemistry for a circular economy. Science 367(6476):369–370, https://doi.org/10.1126/science.aba4979, URL https://www.sciencemag. org/lookup/doi/10.1126/science.aba4979

Lemkowitz SM (1992) A unique program for integrating health, safety, environment and social aspects into undergraduate chemical engineering education. Plant/Operations Progress 11(3):140–150, https://doi.org/10.1002/prsb.720110308, URL http://doi.wiley.com/10.1002/prsb.720110308

Meadows DH, Meadows DL, Randers J, Behrens III, WW (1972) The Limits to Growth, A Report for the Club of romes project on the predicament of mankind. New York: Universe Books. ISBN:0876631650

OECD (2019) Sustainable Chemistry. URL http://www.oecd.org/env/ehs/risk-management/sustainablechemistry.htm

Pearce DW, Barbier E, Markandya A (1990) Sustainable Development: Economics and Environment in the Third World. Earthscan

Primas H (1992) Umdenken in der Naturwissenschaft. GAIA 1(1):5–15

Royal Society of Chemistry (2020) Green Chemistry. URL https://www.rsc.org/journals-books-databases/about-journals/green-chemistry/

Sachverständigenrat für Umweltfragen (1994) Umweltgutachten 1994 - Für eine dauerhaft-umweltgerechte Entwicklung. Tech. rep., URL https://www.umweltrat.de/SharedDocs/Downloads/DE/01_Umweltgutachten/1994_2000/1994_Umweltgutachten_Bundestagsdrucksache.html

Seidl I, Zahrnt A (eds) (2010) Postwachstumsgesellschaft. Metropolis, URL https://www.metropolis-verlag.de/Postwachstumsgesellschaft/811/book.do

Steinemann M, Kohli A, Petry C, von Stokar T (2019) Evaluation der Strategie Nachhaltige Entwicklung 2016-2019. Tech. rep., Infras, URL https://www.infras.ch/media/filer_public/12/0c/120c1117-33d2-443f-bb5a-e917e082b116/evaluation_sne-2016-2019-schlussbericht_are_190710.pdf

Strempel S, Scheringer M, Ng CA, Hungerbühler K (2012) Screening for PBT Chemicals among the "Existing" and "New" Chemicals of the EU. Environmental Science & Technology 46(11):5680–5687, https://doi.org/10.1021/es3002713, URL https://pubs.acs.org/doi/10.1021/es3002713

UBA (2020) Sustainable Chemistry. URL https://www.umweltbundesamt.de/en/topics/chemicals/chemicals-management/sustainable-chemistry#what-is-sustainable-chemistry

UN (1987) Our Common Future. Oxford University Press

UN (1992) Agenda 21. Tech. rep., URL https://sustainabledevelopment.un.org/outcomedocuments/agenda21

UNEP (2019) Global Chemicals Outlook II. URL https://www.unenvironment.org/explore-topics/chemicals-waste/what-we-do/policy-and-governance/global-chemicals-outlook

UNGA (1992) Rio Declaration on Environment and Development. URL https://www.un.org/en/development/desa/population/migration/generalassembly/docs/globalcompact/A_CONF.151_26_Vol.I_Declaration.pdf

UNGA (2015) Transforming our world: the 2030 Agenda for Sustainable Development. Tech. rep., URL https://sustainabledevelopment.un.org/post2015/transformingourworld

US EPA (2020) Green Chemistry. URL https://www.epa.gov/greenchemistry

Wynne B (1992) Uncertainty and environmental learning: Reconceiving science and policy in the preventive paradigm. Global Environmental Change 2(2):111–127, https://doi.org/10.1016/0959-3780(92)90017-2, URL https://linkinghub.elsevier.com/retrieve/pii/0959378092900172

The Role of Legislation

3

3.1 Varied Approaches to Legislation

Ensuring chemical product and process safety to protect human health and the environment is essential, and effective *legislation* containing enforced rules and standards set by governing bodies is central to accomplishing this. The approach taken to develop, enact, and enforce legislation relating to chemical products and processes varies significantly depending on factors such as societal values and previous experiences. Variations in approaches exist between different countries and regions, and differences can even occur on the state and municipal levels.

The chemical industry has an obligation to adhere to the legislation relevant to their operations and products, and this can mean considering multiple, sometimes conflicting requirements when operations and product sales span across borders.

A commonly applied requirement for developing legislation is the ability to establish causality between a damage and its causes. In this regard, the areas of product safety and process safety are very different. While the causes and effects in the area of process safety are usually closely related, dangers posed to human health and the environment by a product can be complex and difficult to identify and assign. This is because the cause and the effect can often be far apart from one another in terms of both time and space (see Fig. 2.3). A wide range of causes can lead to a wide range of health and environmental impacts, some of which may not even be detectable with current techniques.

There are numerous ongoing international discussions regarding which approaches are best to take for developing chemical legislation, including the challenge of determining responsibility for increasingly global chemical pollution problems.

Since process safety legislation has been heavily developed and strictly enforced in many industrialized countries for decades longer than product safety legislation, this chapter focuses on the more recently established legislation for the protection of human health and the environment concerning chemical products, specifically

K. Hungerbühler et al., *Chemical Products and Processes*, https://doi.org/10.1007/978-3-030-62422-4_3

within the European Union (EU). It aims to introduce the fundamentals behind some of the key legislation that exists today and provides links to further resources that describe and discuss it.

Widely seen as currently having one of the most robust and multifaceted sets of chemical legislation, the EU is used as a model for introducing regulatory aspects within this chapter (and also where relevant in later chapters). A shorter introduction is provided for a few other international and national regulatory bodies.

3.2   Developing Measures and Standards

Legislation can be understood as the creation and implementation of binding standards in the common interest. In particular, legislation for safety and protection of human health and the environment deals with the development of a system of rules for handling technology. This concerns, for example, operation, technical design, protection and control measures, as well as questions of liability.

In the EU, five main types of legislation are passed that will be discussed in this chapter (EU, 2019): *Regulations* are binding acts that have to be fully applied as they are written across all EU member states. *Directives* set a goal that all EU member states must achieve, but each member state can establish its own laws to achieve it. *Decisions* are binding acts that must be followed by the countries or entities they specifically address. *Recommendations* are nonbinding and serve as a way for a government entity to make their views on an issue known. Similarly, *opinions* are nonbinding and issued by EU government institutions and committees to clarify their viewpoints.

Developed legislation related to safety and the protection of human health and the environment has two important motives:

1. To react to societally relevant events
2. To proactively align with societal protection goals

The next two sections will elaborate on these two motives.

3.2.1   Reactive Legislation in Response to Relevant Events

Perhaps the most well-known example of reactive product legislation is the ban of the insecticide dichlorodiphenyltrichloroethane, which is commonly known by its acronym DDT. The compound was widely used starting in the 1940s to control arthropods such as lice and mosquitoes to limit the spread of the diseases malaria and typhus. Concerns regarding its adverse effects on the environment were highlighted and brought to the attention of the public in 1962 when Rachel Carson published her now-famous book *Silent Spring* (Carson, 1962). Carson researched and called out strongly suspected links between the increasing use of chemical pesticides and observed damage to human and environmental health. DDT was later linked to causing the thinning of eggshells of wild birds, and the case led to the start of a new environmental movement in the United States. As a result, DDT was banned for agricultural use in the United States in 1972, and in 2004, DDT was listed in the Stockholm Convention on Persistent Organic Pollutant (POPs) and subject to global restriction. It is, however, currently still allowed for use in some regions to control the spreading of insect-borne diseases such as malaria.

As a result of event-related damages linked to man-made chemical technology such as DDT, many new sets of legislation and regulations have been passed through political processes. An example of this reactive process is illustrated in Fig. 3.1. In the figure, a damage to human health or the environment caused by a chemical is communicated via the media to wider society, who then can perceive the risk in different ways. This can lead to a political process that develops legislation in response. This legislation is then enforced and, depending on its effectiveness, should greatly reduce (or hopefully eliminate) any remaining risks that could lead to further harmful accidents or exposures.

Once new legislation has been developed in response to an event and after making its way through an often complex political process, there are still often important challenges to consider during implementation and enforcement:

- Given the number of different environmental regulations that exist, the effectiveness of enforcing them can be difficult to qualify if the legal framework has changed multiple times over the years.
- For some legislation, enforcement can be difficult to do effectively and efficiently given the limited resources available to a government agency.
- Setting new requirements can involve additional investments or de-investments ("divestments") that require time and financial resources. For example, stricter concentration limits in drinking water can require installation of advanced treatment systems, and the restriction of a substance can require a manufacturer to discontinue production and may (temporarily) lay off factory workers employed at a production site.
- Very strict legal limits on emissions could potentially lead to an overall negative environmental impact (see Fig. 3.2), for example, through increased energy demand for the operation of treatment systems.

Today, reactive legislation based on addressing problems as they occur has largely reached its limits. With the current state of environmental law, further progress can likely be achieved more efficiently by consistently focusing on set protection goals. An overview of how this approach might look is provided in the next section.

3.2.2 Proactive Legislation in Alignment with Protection Goals

In a proactive approach to legislation, scientific knowledge and societal consensus are used to set and achieve protection goals. These long-term protection goals

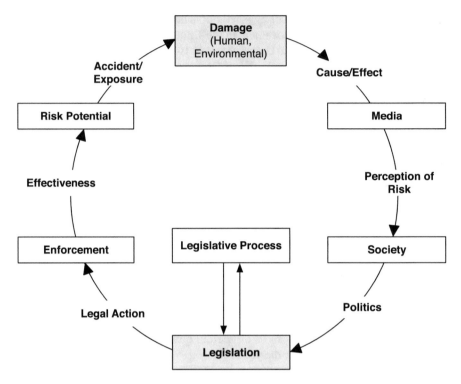

Fig. 3.1 The process of developing reactive legislation in response to a damage

should consider likely future conditions, and they can also be agreed by different countries or regions so that each can enact harmonized regulations to achieve the long-term goals. This also prevents competing companies in different regions from having unfair (and unsafe) advantages. Today, the *precautionary principle* is often explicitly invoked as a proactive basis for legislation. As defined in the United Nations Rio Declaration on Environment and Development (UNGA, 1992) and introduced in Chap. 2, the precautionary principle states that full scientific certainty that a chemical (or chemical product) causes harm cannot be required to take action to prevent such harm.

Environmental strategies based on this approach are already being followed by multiple government entities through incorporating environmental aspects into core legislation. In the EU, proactive legislation is considered beneficial to apply not just in the areas of human health and the environment but generally across all legal domains (European Economic and Social Committee, 2009). A proactive approach for chemical products and processes involves:

- Integration of environmental aspects into overall economic and political considerations.
- Creation of new market incentives instead of mere legal constraints.

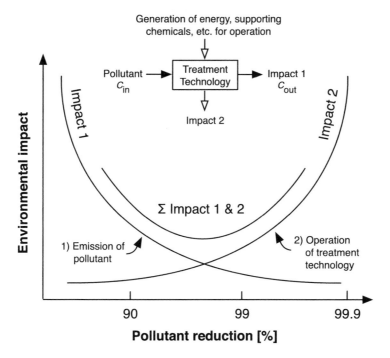

Fig. 3.2 Illustration of the trade-off between treating an emission to reduce the concentration of a pollutant and the impacts caused through the energy demand and production of additional chemicals, etc. required to operate the treatment technology. Past an optimal point, the cumulative curve (in the middle) shows an overall increase in environmental impact associated with further reducing the concentration of the pollutant

- Cooperation between lawmakers, industry, and scientists to achieve maximum synergy between market-based and legally controlled environmental protection. For example, an agreement between industry representatives and authorities regarding voluntary substitution of a hazardous chemical to avoid further state regulations. An example is the Perfluorooctanoic Acid (PFOA) Stewardship Program facilitated by the US EPA (US EPA, 2018a).
- Integrated monitoring of the soil, water, and air as well as biomonitoring of the population.
- Consideration of appropriate limitations on entire groups of similar chemicals (including future potential substitutes) rather than on managing individual chemicals consecutively.
- Requirement of robust evidence if alternatives to hazardous compounds are claimed to not exist or if avoidance and reduction of waste through better process control are claimed not to be possible.

Implementing proactive legislative requires dedication and vigilance among the many individuals and organizations involved in and affected throughout the process, known as *stakeholders*. These can include, for example, legislators, law enforcement agencies, companies, academic researchers, civil society organizations, and the media. In developing and following the legislation, stakeholders should value personal responsibility, precautionary thinking, and market-based optimization. As new technology develops and additional protection options become available, existing standards should also be continuously reviewed and rethought. The various stakeholders are crucial to involve in the legislative development process and, ideally, to also reach a political agreement with. This helps to ensure that legislation receives widespread societal recognition.

3.3   Product Legislation

Globally, numerous existing legislative frameworks relate to chemical products and processes in regard to the protection of human health and the environment. Most often, a country or region (such as the EU) has developed its own set of regulations that address in different ways and to different extents the thousands of chemical substances on the market.

In most places with chemical legislation, newly developed substances need to pass through a defined review process and be approved by government authorities before being allowed on the market. The specific data requirements and criteria for approval can vary greatly between countries and by chemical type.

Given the diversity and complexity of the many regulations and their changes over time, it is not feasible to review and compare all of them here. Instead, this section outlines some of the most important chemical legislations within a few geographic regions with more extensively developed regulations. Focus is placed largely on understanding key legislation within the EU, and a general introduction is provided for legislation in the United States, Canada, Japan, and South Korea. Some key chemical management frameworks and global conventions managed by intergovernmental organizations are also presented.

More information on each of the legislative frameworks or regulations included here can be found by reviewing the websites of the relevant government authorities referenced in each section.

3.3.1   Registration, Evaluation, Authorization, and Restriction of Chemicals (REACH) in the EU

Industrial chemicals within the EU are largely regulated under the regulation known as *REACH*, which stands for the Registration, Evaluation, Authorization, and Restriction of Chemicals (Regulation (EC) No 1907/2007) (ECHA, 2019i). The underlying principle of REACH is that the burden of generating the data needed to prove a substance is safe is placed on the producer or importer of the substance. The required amount of safety data regarding a substance's inherent hazardous properties increases with the increasing amount (tonnage) of the substance annually placed on the market.

Most substances produced in or imported into the EU at an amount of 1 t per year or above must be registered, which requires the collection of a defined set of safety data and submission of a registration dossier to the European Chemicals Agency (ECHA). Member states of the EU, as well as ECHA, can then choose to further evaluate specific substances. Companies wanting to use chemicals that meet set criteria (see Sect. 3.3.1.3) can then be required to have all specific uses of the chemicals authorized in advance by the agency. For dangerous substances, it is possible that their manufacture or placement on the market is specifically regulated or banned under a restriction.

The requirements under REACH do not apply to substances that are radioactive, subject to customs supervision, non-isolated intermediates, wastes, or considered to be included within the transport of dangerous substances. Some of the specific REACH requirements also do not apply to substances that are already regulated under other EU legislation. This includes, for example, pesticides, medicinal products, food additives, and additives in animal feed. The following subsections introduce each of the four main foundations that structure the REACH regulation: registration, evaluation, authorization, and restriction.

3.3.1.1 Registration

Newly manufactured or imported substances need to have a completed *registration* with ECHA before they are allowed to be produced or imported. Substances that were already manufactured or imported before June 2007, when the REACH legislation entered into force, are known as *phase-in* substances. Depending on their production or import volume, these phase-in substances had different deadlines for registration from 2010 to 2018.

To register a substance under REACH, a registration dossier needs to be submitted to ECHA. This always consists of a technical dossier, and if the substance is manufactured or imported in quantities exceeding 10 t per year, a *chemical safety report* (CSR) is also required. The technical dossier needs to include various physicochemical properties of the substance (such as the octanol-water partition coefficient, density, flash point, etc.), as well as toxicological and ecotoxicological information (covering, for example, skin sensitization, short-term toxicity in invertebrates, acute oral toxicity on vertebrates, or biodegradability).

When required, the chemical safety report must show the results of a chemical safety assessment that demonstrates "that the risks from the exposure to a substance, during its manufacture and use, are controlled when specific operational conditions and risk management measures are applied" (ECHA, 2019c). To do this, critical exposure scenarios are defined that describe relevant routes of exposure and releases of the chemical into the environment. This helps to identify any appropriate actions that should be taken to manage risks that might be present. These concepts of exposure and risk management will be more thoroughly introduced and discussed in Chap. 6.

REACH operates under the principle that, conceptually, each substance only needs to have a single registration, even if produced or imported by multiple companies (in reality, however, this is not followed; as of August 2020, there were close to five times as many registration dossiers as substances (ECHA, 2020)). This principle allows registrants to submit a registration dossier jointly and reduce the overall costs by avoiding unnecessary, duplicate testing. To do this, information can be shared between registrants either through a *substance information exchange forum (SIEF)*, using data from preregistered phase-in substances, or through an inquiry to ECHA. ECHA then publishes the information submitted in the registration dossiers that it deemed nonconfidential. All registration dossiers can be accessed on ECHA's website (ECHA, 2020).

Information requirements for registering intermediate substances are reduced, and no chemical safety report is needed. These are defined as substances that exist only temporarily since they are transformed into another substance under controlled conditions and at the same manufacturing site. Substances used in product- and process-oriented research and development (PPORD) can also be exempted from the registration obligation for up to 5 years.

3.3.1.2 Evaluation

Once submitted, registration dossiers can then undergo an *evaluation* by ECHA for completeness. If the dossier involves any proposals for safety testing that makes use of vertebrate animals, these are examined to prevent unnecessary animal testing. In addition, at least 5% of all submitted dossiers in each registration tonnage band (i.e., 1–10 t, 10–100 t, 100–1000 t, and >1000 t) are checked by ECHA to see if they contain all of the required information. Such checks can be either targeted or done randomly, and ECHA has the right to determine if the registrants should submit additional information or carry out additional testing. If ECHA determines this is the case, relevant authorities from EU member states need to also support this decision.

Member states can also themselves request an evaluation of a dossier if there is a concern about the substance's impacts on human health or the environment. This can then result in requesting additional information from the registrant. Should any evaluation conclude that the risks posed by a substance are not being sufficiently managed, then EU-wide risk management measures may be proposed, such as restricting the substance's manufacture, import, or use.

3.3.1.3 Authorization

The process of *authorization* within REACH aims to promote the identification and replacement of highly hazardous substances. Such substances can be formally identified as a *substance of very high concern (SVHC)* following a proposal by an EU member state or ECHA. To do this, the substance has to meet set criteria for either being CMR (carcinogenic, mutagenic, or toxic to reproduction), PBT (persistent, bioaccumulative, and toxic), or vPvB (very persistent and very bioaccumulative). Alternatively, it could also be considered to cause an equivalent level of concern, which has been applied in the case of chemicals with endocrine-disrupting effects (ECHA, 2018).

Proposals to identify an SVHC need to include both information to justify its identification and information on volumes of the substance placed on the EU market, its uses, and possible alternatives. A 45-day public consultation period follows the submission of a proposal, and any stakeholder is invited to provide comments and further information. If the identification is challenged during this process, the case is referred to a ruling board known as the Member State Committee, which then decides unanimously on a verdict or refers the case to the European Commission for a decision. If a substance is successfully identified as an SVHC, it is placed on the so-called Candidate List for Authorization (ECHA, 2019b). This immediately obliges suppliers of the substance to (i) provide customers with a Safety Data Sheet, (ii) communicate on the substance's safe use, (iii) respond to consumer requests for information within 45 days, and (iv) to inform ECHA if any article they produce contains the SVHC in quantities above 0.1% by weight or if they produce or import more than 1 metric ton of the SVHC per year.

Newly identified SVHCs are added to the Candidate List, and ECHA then prioritizes which substances on the Candidate List should be recommended for placement on the Authorization List (ECHA, 2019a). Substances on the Authorization List are prohibited from being used or placed on the market without authorization from ECHA.

To move an SVHC to the Authorization List, ECHA makes a recommendation considering criteria such as inherent hazardous properties of the substance and the scope and scale of use. After a 90-day public consultation period on the recommendation and input from the Member State Committee, the European Commission uses the recommendation to make a final decision. In general, chemicals placed on the Authorization List can be approved for specific applications if an applicant can show that the risks from using the substance are controlled to the point that exposure is below levels not expected to cause adverse effects or if no suitable alternative substances exist to replace it and its socioeconomic benefits outweigh its risks. ECHA's Risk Assessment Committee (RAC) and Socioeconomic Analysis Committee (SEAC) then prepare draft opinions, which provide the basis for the final decision by the European Commission.

3.3.1.4 Restriction

If a substance is found to pose an unacceptable risk(s), a member state or ECHA can suggest applying a *restriction(s)* to the substance, which can include banning it; limiting production, use, or import; as well as requiring specific labeling or technical control measures. Restrictions can also apply to substances where no registration would be necessary, for example, when the production or import volume is below 1 t per year. However, on-site intermediates, substances used in scientific research, and substances that only provide a risk to human health from their use in cosmetics are exempt from any restrictions.

After a dossier proposing a restriction has been submitted to ECHA, a public consultation period begins allowing stakeholders to comment on the proposal. Additionally, and similar to the authorization process, ECHA's RAC and SEAC also give their opinions on whether the proposed restrictions provide adequate measures to reduce the risks and on the socioeconomic impact of the proposed restrictions, respectively. Based on these opinions, the European Commission makes a final decision on the proposed restrictions, which member states then need to enforce.

3.3.2   EU Classification and Labelling (CLP)

Linked to but independent of the REACH regulation is the EU's Classification, Labeling, and Packaging (CLP) Regulation (Regulation (EC) No 1272/2008). It aligns existing European legislation with the United Nation's *Globally Harmonized System of Classification and Labelling of Chemicals (GHS)* (European Commission, 2019a), which is an international standard that aims to do just what its name says: harmonize the global classification and labeling of chemicals. Under CLP, suppliers are responsible for identifying whether a substance or mixture of substances exhibits hazardous properties that might require classification and labeling using the GHS standards. If so, then a GHS hazard category and class need to be assigned to describe and communicate physical, health, and environmental hazards, among others (ECHA, 2019h).

Some of the hazard classifications that exist under the GHS are shown in Table 3.1. Identified hazards for a substance or mixture then need to be communicated to all actors in the supply chain, all the way down to the end consumer. This is done through the use of labels and safety data sheets that aim to alert anyone handling it to the risks, as well as provide information on how to manage those risks. Accordingly, labeling should include the following (ECHA, 2019d):

- Supplier identity: Contact details of the supplier or suppliers have to be featured on the label.
- Product identifiers: The substance(s) should be clearly identifiable through a name as listed in the Classification and Labeling Inventory (see below) and/or through its IUPAC name and its European Community (EC) or Chemical Abstracts Service (CAS) registry number. In the case of mixtures, the trade name of the mixture and the identity of all the substances in the mixture that contribute to its hazardous classification need to be displayed.
- Hazard pictograms: The applicable hazard classification(s) should be displayed in the form of the standardized pictograms, as shown in Table 3.1.
- Signal words: Along with hazard pictograms, appropriate signal words should also be displayed. More severe hazards require the use of the word "Danger" and less severe hazards the word "Warning." In the case of multiple hazard categories, only one signal word should be displayed, with "Danger" taking precedence over "Warning."
- Hazard statements: To describe the nature of the hazard in more details, CLP hazard labels must also include hazard statements. Each statement has a code, starting with an "H" and followed by a three-digit number, indicating the nature of the hazard. Examples can be found in Table 3.2. A full list of current hazard statements is available in the official GHS guidelines (UN, 2019).

- Precautionary statements: Similar to hazard statements, precautionary statements must be included on the label to give more information on how to prevent or minimize adverse effects on human health or the environment. Precautionary statements also have a code, made up of a "P" followed by a three-digit number indicating the category of precaution (see Table 3.2 and UN (2019) for the full list).
- Supplemental labeling information: Under CLP, additional obligatory information to display on the label includes any EU hazard statements (codified as "EUH" statements) and REACH authorization numbers.

To ensure visibility and readability, the CLP Regulation also specifies the size and placement of the label on the package of a chemical product. Manufacturers and importers must also submit the classification and labeling information to the Classification and Labelling (C&L) Inventory of ECHA (ECHA, 2019h). Information on hazardous substances and mixtures also needs to be provided to poison control centers for them to be able to provide rapid medical advice in an emergency.

Only substances and mixtures that are placed on the market fall under the CLP Regulation, which means that it does not apply to non-isolated intermediates or substances only used in scientific research. It also does not apply to radioactive substances. Guidance documentation with further information about complying with the CLP Regulation can be found on ECHA's website (ECHA, 2019h).

3.3.3   Pesticide Legislation in the EU

Developed specifically for their toxic effects on target organisms, *pesticides* can pose unintended yet significant risks to both human health and the environment. To minimize such risks and reduce the overall dependency on the use of pesticides, the EU introduced Directive 2009/128/EC on the sustainable use of pesticides, Regulation (EC) No 1107/2009 on the approval of pesticides, and Regulation (EC) No 396/2005 on the maximum amounts of pesticides allowed to be on or in food and feed (European Commission, 2019c). In these pieces of legislation, the EU follows a twofold approach: First, Directive 2009/128/EC sets out that the dependency on pesticides should be reduced by promoting organic farming and *integrated pest management (IPM)*. The primary aim of IPM is to reduce pesticide use by promoting alternative approaches and techniques to pest control, such as crop rotation, the use of resistant plants, and natural biological and physical methods. If pesticides need to be used, IPM stipulates applying reduced doses with high specificity. According to the regulations, IPM must be implemented by all professional users of pesticides in the EU.

Second, Regulation (EC) No 1107/2009 defines that all *active substances* used in pesticides require approval. It also requires that all final *formulated products* (i.e., containing both the active and inactive ingredients) need to be authorized before use. To seek approval for a novel active substance, manufacturers must apply to an

Table 3.1 Some of the physical hazards, health hazards, and environmental hazards and their respective pictograms according to GHS (UN, 2019)

Physical hazards		
Explosive		• Explosives • Self-reactives • Organic peroxides
Flammable		• Flammables • Pyrophorics • Self-heating • Emits flammable gas • Self-reactives • Organic peroxides • Desensitized explosives
Oxidizing		• Oxidizers
Compressed gas		• Gases under pressure
Physical and health hazards		
Corrosion		• Corrosive to metals • Skin corrosion/irritation • Eye damage
Health hazards		
Toxic		• Acute toxicity (fatal or toxic)
Health hazard		• Carcinogenicity • Mutagenicity • Respiratory sensitization • Reproductive toxicity • Aspiration hazard • Specific target organ toxicity

(continued)

Table 3.1 (continued)

Health and environmental hazards

Harmful		• Acute toxicity (harmful) • Skin and eye irritation • Skin sensitization • Respiratory irritation • Narcotic effects • Hazardous to the ozone layer

Environmental hazards

Environmental hazard		• Acute hazard to aquatic environment • Chronic hazard to aquatic environment

Table 3.2 Examples of some of the hazard statements and precautionary statements according to GHS (UN, 2019)

Hazard statement codes		Examples
H200–H299	Physical hazard	H250: Catches fire spontaneously if exposed to air
H300–H399	Health hazard	H302: Harmful if swallowed
H400–H499	Environmental hazard	H411: Toxic to aquatic life with long-lasting effects
Precautionary statement codes		Examples
P100–P199	General	P102: Keep out of reach of children
P200–P299	Prevention	P231: Handle under inert gas
P300–P399	Response	P375: Fight fire remotely due to the risk of explosion
P400–P499	Storage	P410: Protect from sunlight
P500–P599	Disposal	P501: Dispose of contents/container to ...

EU member state of their choice, which then conducts an initial evaluation. This draft assessment is handed over to the European Food Safety Authority (EFSA), which organizes public and expert consultations before delivering its conclusion to the European Commission where the final decision is made and published. The entire process can take between 2.5 and 3.5 years. Approval of a substance is given for a maximum of 10 years, after which it needs to be renewed. Additionally, the EU has also established a list of active substances that are considered candidates for substitution. For these substances, EU countries need to evaluate whether they can be replaced by other, less harmful chemical or nonchemical alternatives.

Once active substances have been approved, the final pesticide products they are contained within must also be authorized. For this, an application needs to be submitted to and processed by a member state representing one of three zones in the EU (i.e., North, Central, and South). Additionally, other EU countries, the European Commission, and EFSA can also become involved in the application process, which could lead it to take up to 1.5 years in total. In specific cases, existing product authorizations can be expanded, such as adding additional minor uses. In cases where a serious risk needs to be managed, EU member states can also grant emergency authorizations for the controlled use of a product for a maximum period of 120 days.

To protect consumers from exposure to potentially harmful pesticides, Regulation (EC) No 396/2005 sets specific *maximum residue levels (MRLs)*. These values represent the highest legally tolerated concentration of a pesticide residue that is allowed to remain on food or animal feed. MRLs are assessed and set by EFSA considering both the toxicity of each pesticide and different diets so that all demographics, including higher-risk groups such as children, are not exposed at unsafe levels. MRLs are currently set for nearly 300 fresh food products and 1100 pesticides. Where no specific MRL for a pesticide is set, a default value of 0.01 mg/kg is used.

Since many different pesticides are in use, consumers can be exposed to mixtures of pesticides that might result in cumulative effects on their health. This is known as *mixture toxicity*, and research is ongoing to investigate how to adequately consider this when assessing chemical risks. Chapter 6 introduces this topic further and discusses how different chemicals can interact with one another. All active substances and pesticide products currently in use in the EU, as well as their MRLs, can be found online in the EU pesticides database (European Commission, 2019b).

3.3.4 Pharmaceutical Legislation in the EU

Pharmaceuticals are in many ways a very distinctive class of chemicals requiring special attention to protect human health and the environment from unintended adverse effects. In the EU, legislation addressing this topic is mainly set out in Directive 2001/83/EC and Regulation (EC) No 726/2004, both of which are centrally overseen by the European Medicines Agency (EMA) (EMA, 2019b). Pharmaceutical companies according to EU regulation need to apply for authorization from the EMA to market a new drug. Before filing an application, they inform the EMA about their intention to do so, and the EMA offers scientific support in preparing the application to ensure that it complies with the legal and regulatory requirements. This is largely intended to help applicants follow the proper methodology and study design to efficiently generate robust data on the effectiveness and potential risks of the medicine being developed. Potential future patients are also often consulted to identify their needs (EMA, 2019a).

The application dossier needs to include information on the substance itself (including molecular structure, physicochemical properties, stability, biological activity), the drug's effect on patients (including its mode of action, distribution and elimination in the body, and side effects), as well as planned measures to monitor and manage risks after the authorization of the drug. Weighing the benefits and risks of the medicine, the EMA's Committee for Medicinal Products for Human Use (CHMP) will decide on the application within 210 days. During this evaluation period, external experts with special scientific knowledge or clinical expertise are consulted, as well as health-care professionals and patients. Once the CHMP has decided to either authorize or reject the drug, the application is forwarded to the European Commission, where the final, legally binding decision is made and published. After authorization, the EMA's Pharmacovigilance Risk Assessment Committee and EU member states continue to monitor the medicine in terms of its effectiveness, side effects, and other risks that might arise. Should they detect an issue, the committee has the option to restrict the use or suspend the medicine to protect patients (EMA, 2019a).

3.3.5 Key Legislation in the United States and Canada

In North America, both the United States and Canada have set up their own legislation pertaining to chemical products. In the United States, the Environmental Protection Agency (EPA) is mandated to enforce many of the regulations related to the production and use of industrial chemicals. Signed into law in 2016, the Frank R. Lautenberg Chemical Safety for the 21st Century Act (US EPA, 2018b) amends and updates a much earlier chemical legislation known as the Toxic Substances Control Act (TSCA), which had been in force since 1976. The EPA compiles and maintains a list (inventory) of each chemical substance manufactured or processed in the United States, also including imported substances (US EPA, 2018c). Any new substances require advanced submission of a pre-manufacture notice (PMN) that is then reviewed by the EPA to determine if risk management action is needed.

Evaluation of existing substances within the inventory is an ongoing process, and the EPA has been tasked with prioritizing existing substances for evaluation. Substances identified as high priority are then evaluated by the EPA, including hazard and exposure assessments, to determine if the substances present an unreasonable risk to human health or the environment. Notably, the EPA is not allowed to consider non-risk factors (including costs) when making their assessment decision. If a substance is found to pose an unreasonable risk, the EPA must propose a rule that manages the risk. Between its inception in 1976 and the year 2019, however, the EPA had only banned or restricted a handful of chemicals or groups of chemicals through TSCA and the Lautenberg Act (US EPA, 2019).

The management of industrial chemicals under Canadian law is based largely on the Canadian Environmental Protection Act (CEPA), first introduced in 1988 and updated in 1999 (Government of Canada, 2018). Under CEPA, the government is responsible for preventing or reducing risks concerning toxic substances. The definition of substance here includes not only individual industrial compounds but also any mixtures or complex mixtures including those formed naturally and including biotechnology products, effluents, emissions, and wastes. Substances can undergo a risk assessment and then risk management for identified risks to be controlled. An assessment carried out under CEPA aims to determine specifically whether or not a substance is toxic, which is a term defined broadly within the regulation.

Substances already on the market in Canada are included in the Domestic Substances List (DSL), and screening and prioritization efforts have been undertaken to assess some of the more than 20,000 substances included in this list. New substances that are intended to be manufactured in, or imported into, Canada require advanced notification to the authorities, including the submission of a minimum set of data. If the authorities find that the substance could be toxic as defined under CEPA, it could be subjected to restrictions.

3.3.6 Key Legislation in Japan and South Korea

Industrial chemicals in Japan are primarily managed under a set of four main pieces of legislation: the Act on the Evaluation of Chemical Substances and Regulation of their Manufacture (known as the Chemical Substances Control Law (CSCL)), the Industrial Safety and Health Law (ISHL), the Law for Pollutant Release and Transfer Register and Promotion of Chemical Management (PRTR Law), and the Poisonous and Deleterious Substances Control Law (PDSCL). The PDSCL was first introduced in 1950 to protect public health from poisonous and deleterious substances through requiring licenses and standards for manufacturing, importing, and selling substances classified as such (National Institute of Health Sciences, 2018). The ISHL was enacted in 1972 to regulate the manufacture and import of substances to protect human health in the workplace (Japan International Center for Occupational Safety and Health, 2008). It applies bans, required pre-authorizations, or specific labeling requirements to dangerous substances.

First enacted in 1973, the CSCL aims to protect environmental health (Ministry of Economy Trade and Industry, 2018). It requires a pre-manufacture evaluation and approval process for new substances, and existing substances require annual reporting of their manufacturing volumes and use data to authorities if manufactured or imported at levels above 1 t per year. The PRTR Law was introduced in 1999 to manage the reporting of releases and transfers of chemical substances, and it requires certain business types to report these amounts for a specific set of chemical substances (National Institute of Technology and Evaluation, 2018).

In South Korea, industrial chemicals fall under the Act on Registration and Evaluation of Chemical Substances (K-REACH) and the Chemical Control Act (CCA). The K-REACH legislation came into force in 2015 (Korea Legislation Research Institute, 2013) and is known for its similarity to the EU's REACH regulation. It focuses on the registration and evaluation of substances, and it requires importers and manufacturers to register any new substance before being placed on the market as well as any existing substance having a manufacturing volume of 1 t or more per year. As K-REACH was introduced fairly recently, there is a registration grace period allowing importers and manufacturers of existing chemicals to register their substances (depending on the amount) until the year 2030 at the latest. Similar to REACH in the EU, registrations are shared by manufacturers or importers of the same substance, and substances can then undergo a hazard evaluation and risk assessment to determine if any control measures are needed. Substances can also later be prioritized for further assessment, be required to obtain authorization before use, or be restricted or prohibited.

Having entered into force in 2015 in South Korea, the CCA focuses on the reporting of hazardous substances and accident prevention (Korea Legislation Research Institute, 2015). It requires business operators to follow standards for proper handling, storage, and labeling of hazardous chemicals and to create and submit management plans to the proper authorities. The act also requires manufacturers or importers of a chemical to confirm in advance if the substance or any of its ingredients are regulated within South Korea and to then inform the authorities. Restricted substances require permission from the authorities before being imported.

3.3.7   Chemical Management and Frameworks within Intergovernmental Organizations

In addition to national and regional legislation, there is a range of international agreements and frameworks that have been established to address the management of chemicals at the cross-regional and global levels. These can include legally binding *multilateral environmental agreements (MEAs)* such as the Stockholm Convention on POPs as well as voluntary frameworks such as the GHS. An overview of a few of these with references to their official websites is provided in Table 3.3.

3.4   Process Legislation

In addition to legislation that focuses on the management of chemical products, specific legislation also exists for chemical processes to protect human health and the environment. These primarily center around ensuring occupational safety for employees as well as the prevention and management of emissions into the environment. Legislation on process safety is often heavily influenced and motivated by major accidents that have taken place, as is very clearly the case in one of the pieces of legislation introduced in this section. Here, the focus will again be placed on getting familiar with some of the key process-related legislation in the EU.

3.4.1   EU Industrial Emissions Directive

Since 2010, emissions from industrial facilities in the EU have been regulated by Directive 2010/75/EU, which is also referred to as the Industrial Emissions Directive (IED). Under this directive, about 50,000 industrial sites in the EU are required to

Table 3.3 Some of the existing international agreements and frameworks focusing on the management of chemicals and chemical products

Name	Aim/scope	Reference
Basel Convention	Protect human health and the environment against the adverse effects of hazardous wastes	(Basel Convention, 2018)
Rotterdam Convention	Promote shared responsibility and cooperation in the international trade and environmentally sound use of hazardous chemicals	(Rotterdam Convention, 2018)
Stockholm Convention	Protect human health and the environment from persistent organic pollutants (POPs)	(UNEP, 2020)
Minamata Convention	Protect human health and the environment from the adverse effects of mercury	(Minamata Convention on Mercury, 2018)
Montreal Protocol	Reduce or prevent use and release of chemicals that adversely affect the ozone layer (Ozone Depleting Substances, ODS)	(UNEP, 2018b)
Globally Harmonized System of Classification and Labelling of Chemicals (GHS)	International standard (voluntary) created to harmonize the classification and labeling of chemicals and chemical products	(UN, 2019)
Inter-organization Programme for the Sound Management of Chemicals (IOMC)	Group of nine intergovernmental organizations, aims to strengthen international cooperation and increase the effectiveness of the member organizations' chemicals programs	(WHO, 2010)
Strategic approach to international chemicals management (SAICM)	Policy framework (voluntary) to promote chemical safety across sectors and with participation of diverse stakeholders	(UNEP, 2018a)
United Nations Framework Convention on Climate Change (UNFCCC)	International environmental treaty to prevent dangerous human interference with the climate, sets nonbinding limits for greenhouse gas emissions	(UNFCCC, 2019)

obtain a permit and operate under emission limits. The IED is based on five pillars (European Commission, 2019d):

- *Integrated Approach*: The entire environmental impact of a facility needs to be considered, including its emissions to air, water, and land, but also its energy efficiency, generation of waste, accident prevention, generated noise, use of raw materials, and restoration of the site upon closure of the plant.
- *Best Available Techniques (BAT)*: To establish requirements for granting operating permits (such as emission limit values for individual compounds), currently available operating and treatment techniques are reviewed to determine which

should be used as a standard. These BATs and their resulting environmental performances are determined through an exchange of information between experts from the EU member states, industry, and environmental organizations, and this is organized by the European Commission. This results in the creation of BAT reference documents that are then used for setting permit conditions.

- *Flexibility*: In specific cases, EU member state authorities are allowed to set higher, i.e., more lenient, emission limits than those based on the BAT reference documents. This can be the case if the high costs of reaching an emission limit are considered to outweigh the environmental benefits, for example, due to the geographical location, local environmental conditions, or the technical characteristics of the plant.
- *Inspections*: EU member states need to conduct environmental inspections on each site at least every 1–3 years.
- *Public Participation*: For the public to remain informed and able to participate in the decision-making process, permit applications, permits, and emission monitoring results need to be made publicly available.

3.4.2 EU Seveso Directive

Prompted by a catastrophic accident at a chemical plant in Seveso, Italy, in 1976,[1], the Seveso Directive provides legislation to prevent and control major chemical accidents in the EU (European Commission, 2017). In total, there have been three Seveso Directives, with the original from 1982 having been replaced by the Seveso II Directive in 1996 following, among other chemical disasters, the release of methyl isocyanate in Bhopal, India, in 1984. This second version was then later replaced by the Seveso III Directive in 2012 (Directive 2012/18/EU). The Seveso legislation applies to industrial activities where dangerous substances in quantities exceeding a certain threshold are present, but it excludes activities that are covered by other legislation, such as nuclear power plants or transport of dangerous goods. Military sites and extraction and mining operations are also excluded (European Union, 2012).

The core aim of the Seveso Directive is to establish an improvement cycle comprising four elements: prevention, preparation, response, and lesson learning. To achieve this, facility operators and EU member state authorities have several obligations. These obligations follow a tiered approach: Where higher amounts of dangerous substances are present, the industrial facilities are categorized as higher tier and must comply with stricter requirements. Otherwise, facilities are categorized as lower tier, and their requirements are less stringent. Site operators are obliged to notify the authorities if their facility falls under the Seveso Directive. This notification should include information on the substances in use or present at the site (including quantity and physical form), the activity of the installation,

---

[1] See Appendix A for information on this and other historic chemical accidents.

and information on the immediate environmental surroundings of the site (including factors likely to cause or worsen a major accident).

Furthermore, the directive requires operators to establish and implement a *major accident prevention policy (MAPP)*. The MAPP should include not only overall aims and principles of preventive actions but also the role and responsibility of management. It should also show a commitment to continuous improvement. In addition to the MAPP, operators of higher-tier facilities are also required to produce a safety report demonstrating that a MAPP has been implemented, that hazards have been identified with measures taken to prevent major accidents, and that adequate safety and reliability aspects have been taken into account when designing the facility. The MAPP should include detailed information about the organization and its personnel, as well as accident risk analyses and measures of protection and intervention to limit the consequences of a major accident. An internal emergency plan is also mandatory for higher-tier facilities, and this must be provided to the member state authority and include the necessary information to establish external emergency plans.

In the case of an accident, the operator of any site (both higher and lower tiers) is required to inform the competent government authority and provide information on the circumstance of the accident; the dangerous substances involved; available data for assessing the effects of the accident on human health, environment, and property; as well as emergency measures that are being taken. Further, information on how to mitigate the medium- and long-term effects of the accident and how to prevent any recurrence of such an accident needs to be provided.

In addition to defining obligations for facility operators, the Seveso Directive also includes obligations for EU member states and member state authorities. These aim to ensure that land-use policies take into account preventing major accidents and limiting resulting consequences, e.g., through maintaining appropriate safety distances or implementing technical measures. Member states are also required to make any relevant information on sites covered by the Seveso Directive publicly available and ensure that the public is consulted and involved in the decision-making process for individual projects. In the event of a major accident, member state authorities need to ensure that adequate action is taken, that persons likely to be affected are informed, and that the operator takes any necessary corrective measures. Lastly, member states and member state authorities are obliged to conduct inspections and prohibit any unlawful operation of facilities. Any relevant information on the prevention and mitigation of major accidents along with acquired experience should be shared with the European Commission and among other member states.

3.4.3 EU Occupational Safety and Health Directives

The protection of employees while at work is regulated in the EU under the Occupational Safety and Health (OSH) Directives, which are a set of directives managed by the European Agency for Safety and Health at Work (European Union,

2019). The OSH Framework Directive of 1989 (Directive 89/391/EEC) introduced general measures to improve the safety and health of workers throughout the EU. It provided a legal definition of what is considered a work environment, obliged employers to take preventive measures to protect their workers, and introduced risk assessment as a key tool in ensuring workplace safety. In specific regard to exposure to chemicals and to chemical safety, there are multiple, subsequent, and linked OSH directives that are in place. Many of these build on each other and amend earlier directives by including additional substances or addressing further, specific risks. The most important directives related to chemical manufacturing and handling are outlined in this section.

3.4.3.1 Chemical Agents Directive

The most fundamental of the OSH Directives on exposure to chemical agents and chemical safety is the directive on the protection of the health and safety of workers from the risks related to chemical agents at work, also known as the Chemical Agents Directive (CAD) (Directive 98/24/EC) (EU-OSHA, 2019c). It establishes minimum requirements for employers to protect their workers from the effects of chemical agents at their workplace. This includes the employer's obligation to determine whether there are hazardous chemicals present and assess any risk(s) they might pose. In the case of multiple hazardous chemicals, their risks have to be assessed together. Based on these risk assessments, the directive requires the employer to eliminate or reduce the risks to a minimum. This should be done through protection, prevention, and monitoring measures, although substitution of the hazardous substance with a less or nonhazardous one is preferable. In the case of an accident, action plans need to be established that can be rapidly implemented. Furthermore, the directive requires the employer to inform the workers about these emergency procedures, as well as about the results of any risk assessment and the presence of hazardous chemicals at the workplace (including clear labeling of containers and providing access to safety data sheets). Employers are also obliged to train workers on appropriate precautionary and protective measures.

The Chemicals Agents Directive also sets the basis for establishing both legally *binding* and *indicative* (i.e., suggested) external *occupational exposure limits* (OELs) and internal *biological limit values* for hazardous substances. OELs regulate the maximum allowed external exposure of a worker to a substance at the workplace and are derived by considering information on different hazardous properties of a particular substance (ECHA, 2019f). Different routes of exposure can exist, and the directive mainly focuses on exposure via inhalation of vapors, mists, or dust; however, skin penetration is also regulated for certain chemicals (human exposure pathways are further introduced in Chap. 6). Biological limit values rather set maximum allowed limits for workers' internal exposure to a chemical, and compliance with set limits can be determined through biomonitoring techniques (such as measuring concentrations in blood or urine).

EU member states are allowed to set their own *indicative* OELs, and they may be higher or lower than those determined on the EU level. For *binding* OELs, member states still need to set their own national values; however, they are not allowed to exceed values determined on the EU level (EU-OSHA, 2019c). Lists of indicative OELs for several substances are set within four subsequent directives (2000/39/EC, 2006/15/EC, 2009/161/EU, and 2017/164/EU) and can be found online (Institut für Arbeitsschutz der Deutschen Gesetzlichen Unfallversicherung, 2019).

3.4.3.2 Carcinogens and Mutagens Directive

Protection of workers from carcinogenic and mutagenic compounds (excluding radiation) is separately regulated by the EU directive on the protection of workers from the risks related to exposure to carcinogens or mutagens at work, also known as the Carcinogens and Mutagens Directive (CMD) (Directive 2004/37/EC) (EU-OSHA, 2019a). Similar to the Chemical Agents Directive, the CMD requires the employer to assess the risks of substances workers are exposed to at their workplace and take measures to prevent exposure. However, due to the critical nature of carcinogens and mutagens, the CMD has stricter requirements than the CAD. In addition to measures described in the CAD, employers must also keep the number of workers exposed to carcinogenic and/or mutagenic chemicals at a minimum and label and restrict access to areas where such substances are being used. Substance release needs to be minimized, also for emissions into the natural environment. Member states need to establish health surveillance systems for workers, and all cases of occupational cancers should be reported. Records should be kept for at least 40 years after such exposures to help identify long-latency cases. As required by the CAD, all information must be made available to workers.

3.4.3.3 Directive on Exposure to Asbestos

The protection of workers from the highly hazardous minerals known as asbestos is set out in EU Directive 2009/148/EC (EU-OSHA, 2019b). The directive requires that for all activities that might include exposure to asbestos dust, a risk assessment determining the nature and degree of exposure needs to be carried out and include a consultation with the workers. Unless the exposure is strictly sporadic and of low intensity, the employer must report the work activity to the respective EU member state authority. Exposing workers to intentionally added asbestos fibers is generally not allowed, except in the case of demolition or asbestos removal. In these cases, exposure needs to be minimized by using dust-free processes, proper storage, transportation, labeling, and personal protective equipment and also by minimizing the number of workers exposed.

The single maximum limit value of airborne asbestos exposure is set by the directive as 0.1 fibers/cm^3. If this limit is exceeded, the work needs to stop until the cause is identified, appropriate measures are taken, and the effectiveness of these measures is verified. If the limit value cannot be maintained, appropriate personal protection equipment needs to be provided. Furthermore, employers are required to provide their workers with training on asbestos, appropriate protection equipment, and proper waste disposal, among other items. The directive also requires that workers undergo regular health examinations during the exposure and potentially after the exposure. Records on the nature and duration of asbestos exposure need to be kept for at least 40 years by the employer.

3.5   Regulatory Outlook

Legislation addressing chemical products and processes has been constantly developing. The earlier approach of introducing reactive legislation in response to problematic products such as DDT has started to shift toward the introduction of new legislation such as REACH in the EU that aim for a more proactive approach in order to protect human health and the environment before a chemical enters the market.

While this chapter introduces some of the fundamental legislation existing in 2020, changes and adaptations take place regularly. In the EU, assessments of chemicals already on the market as well as chemicals to be introduced into the market will continue for authorization and restriction. New amendments could change the assessment procedures or data requirements, and entirely new pieces of legislation might be developed to address risks posed by future technologies. In the coming years, chemical legislation in Asia, South America, and Africa are especially likely to continue expanding, and they could end up being sometimes very similar to and other times very different from their European and North American counterparts.

In that sense, staying up to date with the latest legislation is an ongoing task, and companies often need to have entire regulatory affair departments to handle the legal obligations for their multiple products that could enter many different international markets. The websites of the regulations, agencies, and organizations referenced in this chapter can serve as initial sources of information for staying informed.

3.6   Room for Improvement

Significant effort has gone into developing and implementing the chemical legislation that exists today. In the EU, significant numbers of stakeholders from government agencies, companies and industry associations, civil society groups, and many others were involved in complex discussions and negotiations over multiple years to make this possible. While this is no doubt a commendable achievement, there should always be room for reflection and continuous improvement. In that spirit, there have been many recent efforts from various stakeholder groups to identify recommendations to increase the protection offered and minimize the administrative burden created by regulations.

For legislation in Europe, a few of the key discussion points and recommendations for improvement include the following:

- Of the REACH registration dossiers that are inspected by authorities, many have been found to be incomplete or inaccurate. (ECHA, 2017; European Environmental Bureau, 2019a,b; Oertel et al., 2018). ECHA has responded through the creation of action plans to increase and improve inspections (ECHA, 2019g) and said that improving compliance is a priority (ECHA, 2019e). The chemical industry has also responded with its own improvement plan (Cefic, 2019).

- Although REACH registration requires the generation of safety data, significant knowledge gaps still exist regarding many of the potential effects that mixtures of chemicals could have on human health and the environment (Kortenkamp and Faust, 2018). Various proposals have been put forward on how to better manage the risks posed by chemical mixtures (Bergman et al., 2019; More et al., 2019; Zilliacus et al., 2019), but no approach has been agreed upon. This aspect will be further discussed in Chap. 6.
- The transparency of the data and processes used in assessing the safety of chemicals within REACH has been scrutinized (Ingre-Khans et al., 2016). Consumers globally have also been concerned with the public availability of safety data on emissions of chemicals into the environment (Client Earth, 2016) and on ingredients used in products (BizNGO, 2019).
- There is a significant ongoing debate regarding the efficiency of the current approach to reviewing the safety of chemicals one at a time. Stakeholders have called instead for the swifter assessment (and regulation) of entire groups of chemicals with similar structures or properties (Cousins et al., 2020; Schneider, 2019).
- There is a call to carefully consider whether or not the use of all chemicals on the market is essential or if instead safer chemical or even nonchemical alternatives can be used to solve current challenges or meet existing needs (Cousins et al., 2019).
- An unprecedented number of research studies and stakeholder initiatives are being carried out on chemicals. All of these data have become scattered across hundreds of different journals and databases (and in many different languages), making it difficult to find and bring all of the relevant information into the legislative decision-making process. Recent calls have been made to better coordinate and engage stakeholders to improve the organization and transfer of these data from scientists to policymakers, as well as to better communicate the policymakers' knowledge gaps back to the scientists (Wang et al., 2019).
- With a rapidly globalizing economy, it is more important than ever to integrate life cycle thinking into chemical legislation. This involves considering the impacts across the entire lifespan of chemical products and processes, from raw material extraction to production and use and through end-of-life treatment. This will be further introduced and discussed in Chap. 5.
- In developing legislation, regulations, and/or agreements, it is important to consider the very different conditions that can exist in different regions. This can be in respect to, for example, climate conditions, existing infrastructure, financial resources, expertise, and safety culture. For example, the specialized safety equipment that might be used (and needed) when applying a hazardous agricultural chemical in a wealthy country may simply not be available (or affordable) to farmers in a developing country.

References

Basel Convention (2018) Basel Convention. URL http://www.basel.int/

Bergman A, Rüegg J, Drakvik E (2019) Final Technical Report EDC-MixRisk: Integrating Epidemiology and Experimental Biology to Improve Risk Assessment of Exposure to Mixtures of Endocrine Disruptive Compounds. Tech. rep., URL https://edcmixrisk.ki.se/wp-content/uploads/sites/34/2019/09/EDC-MixRisk_Final-technical-report_f20190629.pdf

BizNGO (2019) Stakeholder Initiative Seeks Common Ground on Public Policies for Chemical Ingredient Transparency. URL https://www.bizngo.org/resources/entry/stakeholder-initiative-transparency-20190328

Carson R (1962) Silent Spring. Houghton Mifflin

Cefic (2019) REACH Dossier Improvement Action Plan. URL https://cefic.org/our-industry/reach-dossier-improvement-action-plan/

Client Earth (2016) Landmark European court decision grants access to pesticide tests. URL https://www.clientearth.org/landmark-european-court-decision-grants-access-pesticide-tests/

Cousins IT, Goldenman G, Herzke D, Lohmann R, Miller M, Ng CA, Patton S, Scheringer M, Trier X, Vierke L, Wang Z, DeWitt JC (2019) The concept of essential use for determining when uses of PFASs can be phased out. Environmental Science: Processes & Impacts 21(11):1803–1815, https://doi.org/10.1039/C9EM00163H, URL http://xlink.rsc.org/?DOI=C9EM00163H

Cousins IT, DeWitt JC, Glüge J, Goldenman G, Herzke D, Lohmann R, Miller M, Ng CA, Scheringer M, Vierke L, Wang Z (2020) Strategies for grouping per- and polyfluoroalkyl substances (PFAS) to protect human and environmental health. Environmental Science: Processes & Impacts 22(7):1444–1460, DOI 10.1039/D0EM00147C, URL http://xlink.rsc.org/?DOI=D0EM00147C

ECHA (2017) Evaluation under REACH: Progress Report 2017. Tech. rep., URL https://echa.europa.eu/documents/10162/13628/evaluation_under_reach_progress_en.pdf/24c24728-2543-640c-204e-c61c36401048

ECHA (2018) Inclusion of Substances of Very High Concern in the Candidate List for eventual inclusion in Annex XIV: ED/01/2018. Tech. rep., URL https://echa.europa.eu/documents/10162/ede153a4-db00-daf6-120f-6b6ccce0c539

ECHA (2019a) Authorisation List. URL https://echa.europa.eu/authorisation-list

ECHA (2019b) Candidate List of substances of very high concern for Authorisation. URL https://echa.europa.eu/candidate-list-table

ECHA (2019c) Chemical safety report. URL https://echa.europa.eu/regulations/reach/registration/information-requirements/chemical-safety-report

ECHA (2019d) Guidance on labelling and packaging in accordance with Regulation (EC) No 1272/2008. Tech. rep., https://doi.org/10.2823/1132, URL https://echa.europa.eu/documents/10162/23036412/clp_labelling_en.pdf

ECHA (2019e) Improving compliance is ECHA's key priority. URL https://echa.europa.eu/-/improving-compliance-is-echa-s-key-priority

ECHA (2019f) Occupational exposure limits. URL https://echa.europa.eu/oel

ECHA (2019g) REACH Evaluation Joint Action Plan. Tech. rep., URL https://echa.europa.eu/documents/10162/21877836/final_echa_com_reach_evaluation_action_plan_en/0003c9fc-652e-5f0b-90f9-dff9d5371d17

ECHA (2019h) Understanding CLP. URL https://echa.europa.eu/regulations/clp/understanding-clp

ECHA (2019i) Understanding REACH. URL https://echa.europa.eu/regulations/reach/understanding-reach

ECHA (2020) Registered substances. URL https://echa.europa.eu/information-on-chemicals/registered-substances

EMA (2019a) From laboratory to patient: the journey of a centrally authorised medicine. Tech. rep., European Medicines Agency, URL https://www.ema.europa.eu/en/documents/other/laboratory-patient-journey-centrally-authorised-medicine_en.pdf

EMA (2019b) Human regulatory compliance: Overview. URL https://www.ema.europa.eu/en/
    human-medicines-regulatory-information
EU (2019) Regulations, Directives and other acts. URL https://europa.eu/european-union/eu-law/
    legal-acts_en
EU-OSHA (2019a) Directive 2004/37/EC - Carcinogens or mutagens at work. URL https://osha.
    europa.eu/en/legislation/directives/directive-2004-37-ec-carcinogens-or-mutagens-at-work
EU-OSHA (2019b) Directive 2009/148/EC - Exposure to asbestos at work. URL https://osha.
    europa.eu/en/legislation/directives/2009-148-ec-exposure-to-asbestos-at-work
EU-OSHA (2019c) Directive 98/24/EC - Risks related to chemical agents at work. URL https://
    osha.europa.eu/en/legislation/directives/75
European Commission (2017) The Seveso Directive - Summary of requirements. URL http://ec.
    europa.eu/environment/seveso/legislation.htm
European Commission (2019a) Classification and Labelling (CLP/GHS). URL https://ec.europa.
    eu/growth/sectors/chemicals/classification-labelling_en
European Commission (2019b) EU Pesticides Database. URL http://ec.europa.eu/food/plant/
    pesticides/eu-pesticides-database/public/?event=homepage&language=EN
European Commission (2019c) Pesticides. URL https://ec.europa.eu/food/plant/pesticides_en
European Commission (2019d) The Industrial Emissions Directive. URL http://ec.europa.eu/
    environment/industry/stationary/ied/legislation.htm
European Economic and Social Committee (2009) Opinion of the European Economic and Social
    Committee on 'The proactive law approach: a further step towards better regulation at EU
    level'. Tech. rep., European Union, URL https://eur-lex.europa.eu/legal-content/EN/TXT/?uri=
    CELEX:52008IE1905
European Environmental Bureau (2019a) Chemical Evaluation report: Achievements, challenges
    and recommendations after a decade of REACH. Tech. rep., URL https://eeb.org/chemical-
    evaluation-report-achievements-challenges-and-recommendations-after-a-decade-of-reach/
European Environmental Bureau (2019b) Named: major brands 'breaking EU chemical safety
    law'. URL https://eeb.org/named-major-brands-breaking-eu-chemical-safety-law/
European Union (2012) Directive 2012/18/EU of the European Parliament and of the Council
    of 4 July 2012 on the control of major-accident hazards involving dangerous substances,
    amending and subsequently repealing Council Directive 96/82/EC. URL https://eur-lex.europa.
    eu/LexUriServ/LexUriServ.do?uri=OJ:L:2012:197:0001:0037:EN:PDF
European Union (2019) European Agency for Safety and Health at Work. URL https://osha.europa.
    eu/en
Government of Canada (2018) Canadian Environmental Protection Act: general information. URL
    https://www.canada.ca/en/environment-climate-change/services/canadian-environmental-
    protection-act-registry/general-information.html
Ingre-Khans E, Ågerstrand M, Beronius A, Rudén C (2016) Transparency of chemical risk
    assessment data under REACH. Environmental Science: Processes & Impacts 18(12):1508–
    1518, https://doi.org/10.1039/C6EM00389C, URL http://xlink.rsc.org/?DOI=C6EM00389C
Institut für Arbeitsschutz der Deutschen Gesetzlichen Unfallversicherung (2019) Foreign and
    EU limit values. URL https://www.dguv.de/ifa/fachinfos/occupational-exposure-limit-values/
    foreign-and-eu-limit-values/index.jsp
Japan International Center for Occupational Safety and Health (2008) Industrial Safety and Health
    Law. URL http://www.jniosh.go.jp/icpro/jicosh-old/english/law/IndustrialSafetyHealth_Law/
    index.htm
Korea Legislation Research Institute (2013) Act on Registration, Evaluation, etc. of Chemicals.
    URL http://elaw.klri.re.kr/kor_service/lawView.do?hseq=31605&lang=ENG
Korea Legislation Research Institute (2015) Chemical Substances Control Act. URL http://elaw.
    klri.re.kr/eng_service/lawView.do?hseq=34829&lang=ENG
Kortenkamp A, Faust M (2018) Regulate to reduce chemical mixture risk. Science 361(6399):224–
    226, https://doi.org/10.1126/science.aat9219, URL http://www.sciencemag.org/lookup/doi/10.
    1126/science.aat9219

Minamata Convention on Mercury (2018) Minamata Convention on Mercury. URL http://www.mercuryconvention.org/

Ministry of Economy Trade and Industry (2018) CSCL(Chemical Substances Control Law). URL http://www.meti.go.jp/policy/chemical_management/english/cscl/index.html

More SJ, Bampidis V, Benford D, Bennekou SH, Bragard C, Halldorsson TI, Hernández-Jerez AF, Koutsoumanis K, Naegeli H, Schlatter JR, Silano V, Nielsen SS, Schrenk D, Turck D, Younes M, Benfenati E, Castle L, Cedergreen N, Hardy A, Laskowski R, Leblanc JC, Kortenkamp A, Ragas A, Posthuma L, Svendsen C, Solecki R, Testai E, Dujardin B, Kass GE, Manini P, Jeddi MZ, Dorne JC, Hogstrand C (2019) Guidance on harmonised methodologies for human health, animal health and ecological risk assessment of combined exposure to multiple chemicals. EFSA Journal 17(3), https://doi.org/10.2903/j.efsa.2019.5634, URL http://doi.wiley.com/10.2903/j.efsa.2019.5634

National Institute of Health Sciences (2018) Poisonous and Deleterious Substances Control Law. URL http://www.nihs.go.jp/law/dokugeki/edokugeki.html

National Institute of Technology and Evaluation (2018) Information on PRTR SDS. URL https://www.nite.go.jp/en/chem/prtr/prtr_index.html

Oertel A, Maul K, Menz J, Kronsbein AL, Sittner D, Springer A, Müller AK, Herbst U, Schlegel K, Schulte A (2018) REACH compliance: Data availability in REACH registrations Part 2: Evaluation of data waiving and adaptations for chemical 1000 tpa (Final report). Tech. rep., Umwelt Bundesamt, URL https://www.umweltbundesamt.de/en/publikationen/reach-compliance-data-availability-in-reach

Rotterdam Convention (2018) Rotterdam Convention. URL http://www.pic.int/

Schneider J (2019) Urgent action needed on highly persistent PFAS chemicals. Tech. rep., CHEMTrust, URL https://chemtrust.org/pfasbrief/

UN (2019) Globally Harmonized System of Classification and Labelling of Chemicals (GHS), eighth edn. URL https://www.unece.org/trans/danger/publi/ghs/ghs_rev08/08files_e.html

UNEP (2018a) Strategic Approach to International Chemicals Management (SAICM). URL http://www.saicm.org/

UNEP (2018b) UNEP Ozone Secretariat. URL https://ozone.unep.org/

UNEP (2020) Stockholm Convention. URL http://chm.pops.int/

UNFCCC (2019) United Nations Climate Change. URL https://unfccc.int/

UNGA (1992) Rio Declaration on Environment and Development. URL https://www.un.org/en/development/desa/population/migration/generalassembly/docs/globalcompact/A_CONF.151_26_Vol.I_Declaration.pdf

US EPA (2018a) Fact Sheet: 2010/2015 PFOA Stewardship Program. URL https://www.epa.gov/assessing-and-managing-chemicals-under-tsca/fact-sheet-20102015-pfoa-stewardship-program#what

US EPA (2018b) The Frank R. Lautenberg Chemical Safety for the 21st Century Act. URL https://www.epa.gov/assessing-and-managing-chemicals-under-tsca/frank-r-lautenberg-chemical-safety-21st-century-act

US EPA (2018c) TSCA Chemical Substance Inventory. URL https://www.epa.gov/tsca-inventory

US EPA (2019) Regulation of Chemicals under Section 6(a) of the Toxic Substances Control Act. URL https://www.epa.gov/assessing-and-managing-chemicals-under-tsca/regulation-chemicals-under-section-6a-toxic-substances

Wang Z, Summerson I, Lai A, Boucher JM, Scheringer M (2019) Strengthening the science-policy interface in international chemicals governance: A mapping and gap analysis. Tech. rep., International Panel on Chemical Pollution, https://doi.org/10.5281/zenodo.2559189, URL https://www.ipcp.ch/wp-content/uploads/2019/02/IPCP-Sci-Pol-Report2019.pdf

WHO (2010) The Inter-Organization Programme for the Sound Management of Chemicals (IOMC). URL http://www.who.int/iomc/en/

Zilliacus J, Beronius A, Hanberg A, Luijten M, van Klaveren J, van der Voet H (2019) Deliverable 8.3: EuroMix handbook for mixture risk assessment https://doi.org/10.5281/zenodo.3560720, URL https://zenodo.org/record/3560720

Placing Integrated Development into the Business Perspective

4

4.1 Chemistry: High Performance Combined with a Wide Range of Risks

Both the production and use of chemical products are associated with a wide range of risks, and integrating safety and the protection of human health and the environment into the business model of the chemical industry is not only sensible (and often legally mandatory as seen in Chap. 3), but it also adds value.

Specialty chemistry is a particularly good example. Specialty chemicals are active ingredients, materials, and fine chemicals with a high degree of originality, quality, and value, and they require significant investments in research and development. The market leaders in specialty fields are often responsible for creating the greatest added value in chemistry, and speed during innovation is crucial given that patents are only valid for a set number of years, and manufacturers need to sell and make a profit while they have exclusive rights to the products in their markets. While successful development of a new substance can lead to incredible market success, a single manufacturing accident or an undetected toxic effect can erase all of the value generated.

In chemical production, substances are often used under demanding chemical and physical conditions, and the substances themselves can be highly reactive, toxic, corrosive, flammable or explosive. Their production, storage, and transport can take place in processes and facilities prone to technical faults, and human errors can lead to accidents. Even without accidents, pollution can still occur through waste generation and emissions, and adverse side effects from product use can seriously damage humans and ecosystems. With chemical products being more widely used in technical applications and everyday consumer products in an ever greater variety and quantity, the amount and complexity of potential risks are increasing and need to be considered carefully. The perception of the public has also become more important than ever before. Chemistry today is part of a society whose prosperity is

K. Hungerbühler et al., *Chemical Products and Processes*,
https://doi.org/10.1007/978-3-030-62422-4_4

closely linked to the value created by the chemical industry, but it is also operating within a society that has become increasingly critical of chemical risks.

This chapter aims to introduce the evolution of integrated development in the chemical industry, the value it provides from the business perspective, and the three guiding principles that define it.

4.2   The Ongoing History of Integrated Development

Until the mid-twentieth century, integrated development in the chemical industry was hardly considered systematically. Historically speaking, it first began through a focus on evaluating the safety of production processes and then expanded and evolved to also include environmental pollution and resource uses:

- *Safety:* Beginning in the 1960s, growing awareness of safety concerns, increasing production volumes, and more complex technologies led to the systematic investigation of occupational hygiene, thermal process safety, and explosion prevention technology. Scientific findings within industry resulted in setting new safety standards that were anchored in using risk assessment for production plants, processes, and products.
- *Emissions:* Advancement in the control of factory emissions took place just a few years later, and the relevance of emissions into the water, air, and soil compartments were recognized. The establishment of new environmental measurement methods and the use of *end-of-pipe* treatment technology (e.g. wastewater treatment plants, exhaust-air purification systems, waste incineration plants) also played a key role.
- *Energy:* Initially mostly seen as a cost component and recognized later as an important aspect of environmental protection, energy consumption has been subject to frequent optimization since the oil crisis of 1973.
- *Environmental management:* More careful consideration of emissions, resource use, and waste then followed. Since the 1980s, aspects of integrated development have increasingly found their way into corporate principles and management processes (see also Kostka and Hassan (1997)), often under the umbrella of 'environmental management' or 'Environment, Health, and Safety (EHS)'.

This expansion over time was driven also by criticism from the general public. The perception of risks by a professional risk assessor can be very different from the perception of a lay community member living across the street from a large chemical manufacturer. Society's critical stance toward the chemical industry came about through not only a fundamental change in values but also from specific chemical-related incidents. Some of the multiple and serious chemical accidents include those in Seveso (in 1976), Bhopal (in 1984) and Schweizerhalle (in 1986) (see Appendix A). These were all very visible and high-impact events directly resulting from chemical production. Large communities were affected through loss of life or property as well as damage to the surrounding environment.

In addition to manufacturing accidents, damaging, long-term emissions from normal manufacturing operations have also gained public attention. Perhaps the most famous example is the emission of methylmercury from the Chisso Corporation in Minamata, Japan, which began in the early 1900s and led to thousands of local deaths and permanent damage to human and environmental health (Harada, 1995). More recently, the leaching of perfluorooctanoic acid (PFOA) into drinking water supplies in the United States from DuPont-owned sites was discovered. In 2017, lawsuits for human health damages over the pollution were settled (Reisch, 2017), and in 2019 a major Hollywood film was released dramatizing the legal battle (IMDb, 2019) and bringing even more global public awareness to the issue. Such events often led to a rapid tightening of local safety and environmental legislation through political processes, many in the 1980s. The chemical industry responded using various measures and sought improvements in their technical and scientific understanding of operations as well as in new management processes:

- Construction of expensive end-of-pipe treatment and safety infrastructure
- Improvements in gathering core data to describe their operations, for example, through the development of environmental measurement methods and process- or plant-specific material and energy balances
- Completing and expanding risk assessments of products and processes
- Expanding company safety and environmental protection culture

Separate from manufacturing operations and accidents, awareness has more recently grown regarding the risks of chemical products in causing damage to human health and the environment. This awareness has been made possible largely through the recent advancements in environmental analytics (e.g. trace analysis of chemicals in the environment, biomonitoring of humans, etc.), which have for the first time made the fate and widespread global presence of many synthetic chemicals in the environment visible.

Ongoing concerns have been expressed by various civil society groups as well as academic scientists regarding topics such as the widespread use of pesticides in agriculture (Reuters, 2018), pollution of the oceans by microplastics (Worm et al., 2017), and use of chemicals that can interact with (and disrupt) the endocrine system in everyday consumer products (Dodson et al., 2012). While progress continues to be made to formally address some of these critical chemical issues (such as the *Minamata Convention on Mercury* that came into force in 2017 (Minamata Convention on Mercury, 2018)), there are still many ongoing discussions, unresolved issues, and new concerns to address.

Despite improvements that have been made in many regards over the last few decades, the use of integrated development in the chemical industry is clearly still needed today. Considering the larger variety of products on the market and the ever-increasing global annual production volume of chemicals, perhaps it is more important now than ever before. Reactive approaches such as end-of-pipe treatments have largely reached their limits, and instead, a proactive approach is needed to prevent these emissions (and their impacts) from the very beginning.

The next section introduces these two approaches as well as the guiding principles of integrated development and management systems that can be used to help implement it.

4.3 Integrated Development as Part of the Business Strategy

4.3.1 Reactive and Proactive Approaches

Just as introduced in Chap. 3 for legislative processes, two types of approaches can be taken within the chemical industry to ensure safety and the protection of human health and the environment. A *reactive* approach is characterized by symptom control. Following an accident or discovery of damages from a product, for example, a company (or legislator) can create new standards in response to prevent this from happening again in the future. This approach only addresses problems after they have occurred. Aside from the health and environmental damage caused, this can also result in multiple (and expensive) fines and lawsuits. Furthermore, it is clear that society will have little confidence in the responsibility of the chemical industry under this approach.

In contrast, the approach of the chemical industry could instead be one that is *proactive*. In this case, clear and ambitious goals are set for ongoing improvement, and the industry takes responsibility for implementing changes to achieve them. This creates an internal incentive for safer, higher-quality products and processes. While working toward set goals, the industry can also communicate their efforts to the broader public, and legislators can use the industry's expertise to develop protection-oriented legislation.

Being proactive in the development and use of safe chemical processes and products should be seen as advantageous rather than as a limitation. A proactive approach avoids problems before they occur and builds trust within society. Compared to just a few decades ago, the general public is much more aware and critical of the presence of chemicals in products they buy. A proactive mindset helps the chemical industry to be (and be seen as) part of the solution rather than as a source of the problem.

4.3.2 Guiding Principles of Integrated Development

The concept of integrated development is based on the three guiding principles of eco-efficiency, inherent safety, and social acceptability. These help to (1) optimize the efficiency of resource use and limit emissions, (2) consider proactive measures and minimize hazards, and (3) maximize the potential for acceptance by relevant stakeholders. Implementation of each guiding principle is supported by one of the three methods introduced in the next part of this book (life cycle assessment, risk assessment, and risk-benefit dialogue). Figure 4.1 illustrates how the guiding principles and methods jointly regulate the iterative development needed to create

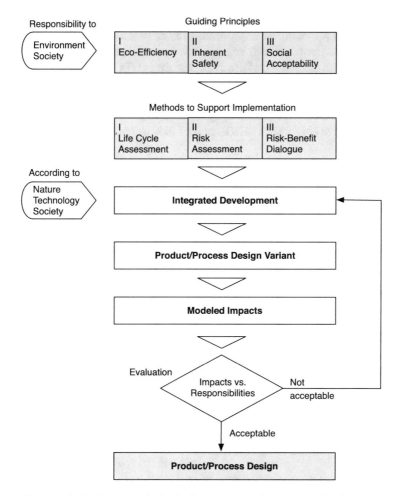

Fig. 4.1 Integrated development of chemical processes and products with the three guiding principles and the methods and tools used to implement them. An iterative approach evaluates design variants and compares their impacts to the responsibility to protect human health (society) and the environment

high-quality chemical processes and products across their entire life cycle. These should be applied to both chemical processes as well as the chemical products that these processes help to manufacture. A well-developed process that follows these principles does not compensate for the production of a poorly-developed product that does not follow the principles.

4.3.2.1 Eco-Efficiency
The principle of eco-efficiency relates the amount of value added from a product or process to the required amount of resources used or polluted to create it. High eco-

efficiency is reached when a product or process is competitively priced, satisfies human needs by providing an improved quality of life, and at the same time limits environmental and human health impacts. The eco-efficiency of a process or product is determined with the method of life cycle assessment (LCA) introduced in Chap. 5.

Principle of Eco-Efficiency Proactively minimize resource use and environmental impacts per unit of service provided by a chemical product or process rather than relying on reactive approaches such as end-of-pipe technologies.

4.3.2.2 Inherent Safety

Inherent safety is a characteristic of a product or process that by design limits any hazards to a predetermined level deemed acceptable. This includes, for example, that highly toxic, flammable, or aggressive chemicals are not used at all in a certain process. The level of inherent safety depends on the size of the remaining hazard potential and is determined using the methods of product risk assessment (Chap. 6) as well as process risk assessment (Chap. 7) and thermal safety (Chap. 8).

Principle of Inherent Safety Proactively limit hazards rather than use reactive risk management.

4.3.2.3 Social Acceptability

Social acceptability can be understood as a society's consensual acceptance of a risk. Risks are accepted when the expected benefit is thought to be greater than the potential damage from a chemical product or process. Examples of expected benefits include the function provided by the product or process itself, as well as positive economic effects on entire regions (e.g. employment, tax revenues), or the enabling of new technologies. Social acceptability is achieved through societal dialogue about risks and benefits (Chap. 9).

Principle of Social Acceptability Proactively ensure that a societal dialogue on the risks and benefits takes place rather than relying on reactive management of societal conflicts.

4.3.3   Integrated Management Systems

A proactive approach implements integrated development as completely as possible into business processes. This is done through an *integrated management system* for continuous (iterative) improvement of business performance. Figure 4.2 provides an illustration of this. This type of system should be implemented within a company's existing business processes in order to apply integrated development at three levels: the first sets normative business principles, the second develops strategic long-term goals, and the third aligns these with day-to-day tasks at the operational level.

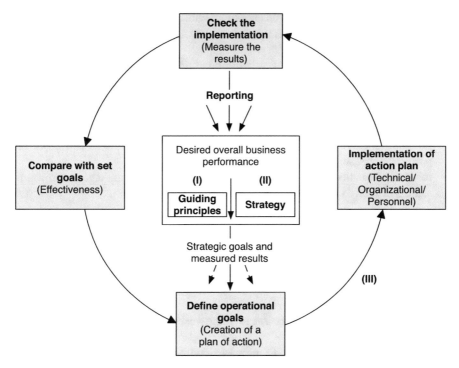

Fig. 4.2 Integrated management system for the continuous improvement of business performance regarding the three guiding principles of integrated development. (I) Normative level (corporate principles); (II) Strategic level (long term goals); (III) Operational level (instructions for operational implementation)

4.3.3.1 Normative Level (I)

At the normative level, integrated development is first anchored into corporate principles through creation of a mission statement for the protection of human health and the environment. Values are defined and declared that the company then upholds for itself and also for its stakeholder groups (e.g. customers and suppliers, employees, investors, the public, etc.). This normative level represents a commitment from top management to apply integrated development and, in doing so, to include all employees, to cooperate with customers and suppliers, and to maintain open communication with the public.

4.3.3.2 Strategic Level (II)

At the strategic level, the mission statement is translated into long-term goals for the company's values, business units, markets, and technologies. Strategic goals can include, for example, the achievement of:

- the application of integrated development within individual business processes, including the creation of clearly defined tasks, competencies, and support by specialists
- the institutional anchoring of suitable conditions and methods for integrated development; examples are the principle of product stewardship[1] and tools such as life cycle assessment, risk assessment, and risk-benefit dialogue with society
- an information and control system to continuously measure the efficiency and success of integrated development efforts; it is helpful to quantify (to the extent possible) the related risks, damages, and burdens, as well as the costs for protection measures
- the use of quality and environmental management systems, e.g. according to standards such as ISO 9000 and ISO 14000

4.3.3.3 Operational Level (III)

At the operational level, objectives are defined for the decentralized implementation of the strategic goals in the individual business and project organizations. At this stage, a broad implementation of integrated development and the relevant management system takes place with the following elements:

- development of the organization to improve business processes in line with strategic goals
- development of employees to promote competences through corporate co-responsibility as well as ongoing education and training
- development of improved technology, to have increasingly environmentally responsible products and processes
- development of internal and external communication to promote acceptance within society through ongoing dialogue with stakeholders
- monitoring progress to make business processes more effective and efficient through regular comparisons with targets

To ensure that a management system is running effectively, it is important to check and communicate its performance. Today this is commonly done through the publication of an annual safety and environmental report that includes the set targets as well as aggregated performance indicators as proof of the progress made toward achieving them.

---

[1] Product stewardship: Responsible product management that spans the entire life cycle of a product, including especially use, recycling and disposal.

4.4   The Importance of Starting Early

It is advantageous to employ integrated development as early as possible. As Fig. 4.3 illustrates and Heinzle and Hungerbühler (1997) explain, the freedom of design decreases quickly during the development of a product or process. This means that as much information as possible regarding design alternatives should be collected as early as possible, and this knowledge should be systematically considered to make informed decisions, despite time pressure and a potentially limited set of initial data. The flexibility to make adjustments in the early stages of development should be used to make decisions that provide benefits in the long term. Once, for example, an entire production plant has been designed to produce a substance requiring a specific production process, making additional changes to the chemical structure of this substance will be a much more difficult and time-intensive task. To overcome the challenge of efficiently integrating all of the relevant information during development, it is important to:

- initially evaluate a broad variety of design variants
- always keep in mind the intended overall development process and its necessary conditions when focusing on areas specific to an individual stage of development
- undertake an increasingly in-depth assessment of design variants that always takes into account the three guiding principles as well as costs
- continuously collect information to develop a knowledge base that can support decision making along the entire development path
- allow for people with different backgrounds (marketing, research & development, engineering, production, EHS, etc.) to communicate with one another in flexible teams
- ensure good cross-flow of information between parallel development processes

Placing integrated development into the business perspective is a benefit for the chemical industry. Considering the three guiding principles helps to strengthen a company's bottom line and to meet the responsibility that the company has to protect human health and the environment. The next chapter begins Part II of this book, which introduces the three key methods of LCA, risk assessment, and risk-benefit dialogue.

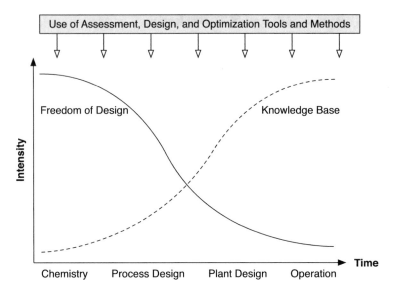

Fig. 4.3 The availability of knowledge and freedom of design over time during development of a chemical production process

References

Dodson RE, Nishioka M, Standley LJ, Perovich LJ, Brody JG, Rudel RA (2012) Endocrine Disruptors and Asthma-Associated Chemicals in Consumer Products. Environmental Health Perspectives 120(7):935–943, https://doi.org/10.1289/ehp.1104052, URL http://ehp.niehs.nih.gov/1104052

Harada M (1995) Minamata Disease: Methylmercury Poisoning in Japan Caused by Environmental Pollution. Critical Reviews in Toxicology 25(1):1–24, https://doi.org/10.3109/10408449509089885, URL http://www.tandfonline.com/doi/full/10.3109/10408449509089885

Heinzle E, Hungerbühler K (1997) Integrated Process Development: The Key to Future Production of Chemicals. URL https://www.ingentaconnect.com/content/scs/chimia/1997/00000051/00000005/art00003

IMDb (2019) Dark Waters. URL https://www.imdb.com/title/tt9071322/

Kostka S, Hassan A (1997) Umweltmanagementsysteme in der chemischen Industrie. Springer Berlin Heidelberg, Berlin, Heidelberg, https://doi.org/10.1007/978-3-642-59057-3, URL http://link.springer.com/10.1007/978-3-642-59057-3

Minamata Convention on Mercury (2018) Minamata Convention on Mercury. URL http://www.mercuryconvention.org/

Reisch MS (2017) DuPont and Chemours settle PFOA suits. URL https://cen.acs.org/articles/95/web/2017/02/DuPont-Chemours-settle-PFOA-suits.html

Reuters (2018) Bayer's Monsanto faces 8,000 lawsuits on glyphosate. URL https://www.reuters.com/article/us-bayer-glyphosate-lawsuits/bayers-monsanto-faces-8000-lawsuits-on-glyphosate-idUSKCN1L81J0

Worm B, Lotze HK, Jubinville I, Wilcox C, Jambeck J (2017) Plastic as a Persistent Marine Pollutant. Annual Review of Environment and Resources 42(1):1–26, https://doi.org/10.1146/annurev-environ-102016-060700, URL http://www.annualreviews.org/doi/10.1146/annurev-environ-102016-060700

Part II

Technical Foundation

Life Cycle Assessment of Chemical Products and Processes

5

5.1 The Importance of Considering All Impacts

To responsibly innovate and design high-quality chemical products and processes, the economic, environmental, and societal impacts they cause need to be carefully considered. While investment calculations have been around for decades to help with the economic analysis of a project, environmental and societal impacts, however, have classically been ignored as *external costs* and not considered in the design of chemical products and processes. As introduced in Chap. 2, external costs are those incurred and imposed onto a third party by a producer of a good or service. In the context of protecting human health and the environment, these can include costs resulting from, for example, public waters being polluted by a factory's emissions.

The importance of considering these impacts has been increasingly recognized through the development and application of a method called *life cycle assessment* (LCA) for health and environmental impacts and *social life cycle assessment* (SLCA) for social impacts. The scope of traditional investment calculations is also now being extended in order to consider costs along the entire life cycle of a product, and this is implemented through a method known as *life cycle costing* (LCC).

The three of these methods together make up the conceptual framework for an entire *life cycle sustainability assessment* (LCSA): LCSA = LCA + SLCA + LCC (Swarr et al., 2011a). So far, LCA is the only one of these three methods that has been internationally standardized through the International Organization for Standardization (ISO) 14040 (ISO, 2006a) and 14044 (ISO, 2006b) standards.

This chapter focuses on introducing the basic framework specifically surrounding LCA for understanding the health and environmental impacts of chemical products and processes. Its role in implementing the first guiding principle of *eco-efficiency* is presented, and a brief introduction is also provided for key elements of LCC and SLCA.

5.2   Introducing Life Cycle Assessment

While economic accounting and valuation of a product or a process can usually be based on pricing in existing markets, accounting for health and environmental impacts raises fundamental questions about which natural resources to protect and how to prioritize them. Since every man-made system within the *technosphere* is fundamentally in conflict with the natural systems of the *ecosphere* (also referred to as the biosphere), it is important to find measurements that enable an evaluation of the impacts and damages that can occur. With a wide range of very different types of potential damages, there also needs to be a way to compare damages with one another.

Life cycle assessment aims to help systematically overcome these challenges. It does this by identifying, quantifying, and interpreting the impacts caused by resource consumption and emissions as completely as possible while remaining objective, transparent, and practical.

LCA is an appropriate method to support:

- Grassroots design: prioritization of research and development (R&D) projects for new products or processes
- Retrofit design: prioritization of R&D projects for existing products or processes
- Investment projects and supply chain management: selecting among available technologies, investigating steps within a supply chain, considering the location of suppliers, etc.
- Policy and marketing strategies: taking into account retailers and customers, providing information to consumers, supporting legislative decisions, etc.

LCA is limited in the support it can provide for:

- Consideration of risks and impacts from the use of a chemical product. These can only be relatively assessed by LCA but can be more specifically characterized through *product risk assessment* (see Chap. 6).
- Consideration of accident-related impacts. This is better addressed through *process risk assessment* (see Chap. 7).

While the basic approach and steps of completing an LCA have largely remained the same since its first standardization in the 1990s, various methods for calculating impacts have continued to develop. Much-needed data describing the resource use and interactions that a product or process has with the environment have significantly expanded (Hellweg and Milà i Canals, 2014).

An LCA is subdivided into four main phases following the ISO standard 14040 (ISO, 2006a). These include (1) *goal and scope definition*, (2) *inventory analysis*, (3) *impact assessment*, and (4) *interpretation* (see Fig. 5.1). Each of these phases is introduced and discussed in more detail in this chapter. For further, in-depth explanations of each phase and additional examples of applying LCA, see Jolliet et al. (2015), Hauschild et al. (2018), and ISO (2006b).

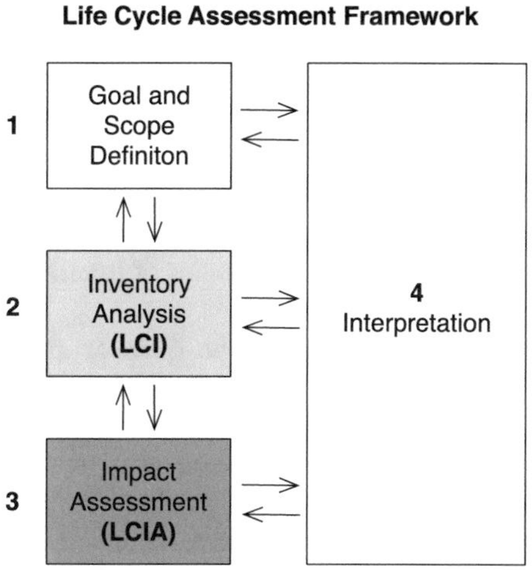

Fig. 5.1 The four phases of life cycle assessment according to ISO 14040 and the flow of information between them

Life Cycle Assessment Framework

5.3   Goal and Scope Definition (Phase 1)

The starting point of an LCA is to first define the *goal* of completing the assessment. This includes defining what particular questions an LCA should help answer, the intended purpose of completing the LCA, and the target audience of the results. In addition, the *scope* of the assessment should be defined and include specifying the *functional unit* that will be used to compare alternatives as well as the *system boundaries* and the width and depth of the analysis. This is the more qualitative phase of LCA that provides the context and structure needed for moving forward with data collection and impact calculations in the later phases.

Concerning the integrated development of chemical products and processes, three typical aims are often the motivation behind completing an LCA:

- To assess the impacts across the life cycle of a new product or process resulting from its design
- To compare existing products or processes
- To improve an existing product or process

Typical questions that can be answered in an LCA also concern better understanding the relative importance of different life cycle stages of a product and to help compare individual impact categories within the total impact. Depending on the goal, the requirements for the scope and accuracy of an LCA can vary. In development projects, for example, LCA can serve as an effective decision-making method at early stages. Even with limited data, an LCA based on key indicators such as energy and material use, water consumption, solvent consumption, production of coproducts and waste, critical individual substances, etc. can provide important screening information on the environmental compatibility of product and process alternatives.

The *functional unit* is an important aspect within an LCA. It can be understood as the reference utility or service to which input and output flows and the associated impacts of a product or process are all related. It should express the benefits of the technical system (normally a product or process) under review as comprehensively and quantitatively as possible, which means it should be clearly definable and measurable. A few examples of functional units include:

- For an herbicide: The quantity of the herbicide required for the control of infestation by weeds of a given cultivated area with a specific selectivity and over a certain period
- For a textile dye: The quantity of dyestuff required to color a given amount and quality of cotton in a specific way (considering desired color, lightfastness, etc.)
- For the production of a chemical: Production of a specific quantity (e.g., 1 kg) of a given chemical of a certain quality that has to be produced under specific technical and socioeconomic conditions

The *system boundaries* that need to be defined within an LCA include the spatial and temporal limits of the system, processes and emissions to be neglected, as well as the definition of *allocation* criteria (introduced later in Sect. 5.4.2). The desired *impact categories* to be calculated are also chosen and documented in this first step, along with the methods and models that will be used to calculate them. These are introduced and discussed in more detail in Sect. 5.5.

Naturally, the results of a life cycle assessment are associated with uncertainties. When defining the goal and scope, it is important to also determine the *width and depth* needed for the analysis. This includes the completeness, consistency, and level of data quality required, as well as how data gaps will be treated. The uncertainty can be characterized using confidence intervals and sensitivity analyses or by error calculations. Recent literature has provided a better understanding of how system assumptions and value choices can impact overall uncertainty (De Schryver et al., 2013; Gregory et al., 2013), and quantitative uncertainty assessment methods are also available (Lloyd and Ries, 2008).

The set goal and scope of the LCA also have an impact on the choice of background technologies (such as energy supply and waste disposal methods) that will be used in the second LCA step of *inventory analysis*, which is introduced in the next section. For example, while the evaluation of existing products and processes may be based on averaged data from existing technologies, development projects should use background data from the latest or even future (prospective) technologies, such as advanced renewable energy sources or recycling processes.

5.4   Inventory Analysis (Phase 2)

The second phase of LCA involves the creation of the *life cycle inventory* (LCI), which quantifies the various mass and energy flows involved, with the focus placed on the interface between the technosphere and ecosphere (see Fig. 5.2). The outcome of the LCI phase is a list of resource uses and emissions. Experience has shown that the preparation of the life cycle inventory is the most time-consuming step in a life cycle assessment.

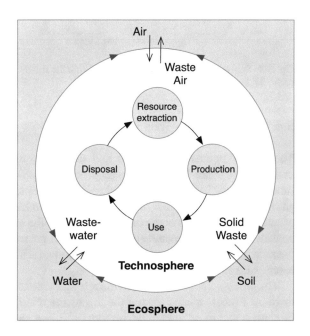

Fig. 5.2 Quantifying the flows between the technosphere and ecosphere is the focus of the life cycle inventory analysis (LCI) phase of an LCA

Flows that leave the system boundaries as outputs into the economy such as the product and coproducts are known as *product flows*. Within the system boundaries, a *unit process* is the smallest element considered in the LCI for which input and output flows are quantified. Flows between these unit processes are known as *intermediary flows*. Flows of natural resources extracted from the environment or of emissions emitted by the system into the environment are known as *elementary flows*, and these are later specifically assessed in the next phase of life cycle impact assessment (LCIA).

Figure 5.3 shows the structure of this system, which includes the elementary flows of the material and energy resources. The product flows leaving the system include the functional unit (which is often the product under review) and a set of by-products.

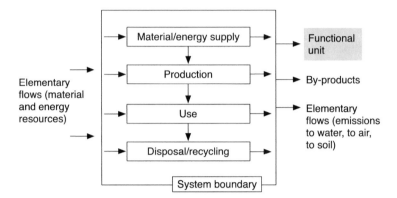

Fig. 5.3 Schematic of the flows considered within the life cycle inventory (LCI) of a product life cycle. All flows are related to a functional unit

5.4.1 The Process Tree

The starting point for identifying and structuring the flows within an LCI is the main process chain. This includes all process steps which are directly related to the functional unit. To illustrate this, a simplified example for the production of chlorine gas is used here. Figure 5.4 shows the process tree leading to the production of 1 kg of chlorine gas, which is the functional unit. In the case of this example, processes considered in the tree include the production of the necessary salt and electricity for the reaction, as well as the steel needed to manufacture the equipment in which the reaction takes place.

The process tree contains all of the relevant secondary process chains as branches or roots. Process chains toward the top of the tree ($n = 1$) are called *foreground processes*, and those toward the bottom ($n = 2$, $n = 3$, etc.) are called *background processes*. Note that in addition to the generation of the desired chlorine gas, this reaction also produces the coproducts of hydrogen gas (H_2) and sodium hydroxide (NaOH). Coproducts can be considered specifically within an LCA through an approach introduced in the next section.

5.4.2 Allocation

In addition to the main products being accounted for during an LCA, many systems also produce additional coproducts. It is important that the resource uses and emissions of the system are all distributed among these multiple products. This distribution is referred to as *allocation*. To do this, the resource uses and emissions need to be inventoried according to a defined allocation method, which should be decided upon and documented at the beginning of the LCA and then followed throughout.

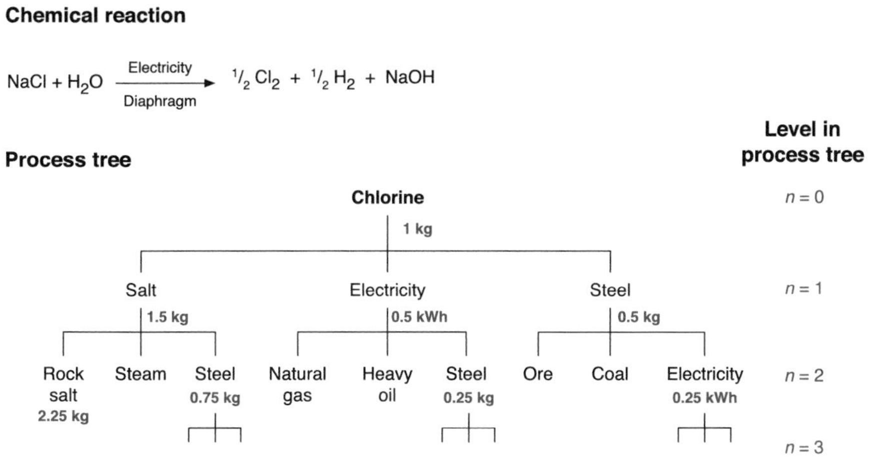

Chemical reaction

$$NaCl + H_2O \xrightarrow[\text{Diaphragm}]{\text{Electricity}} \tfrac{1}{2} Cl_2 + \tfrac{1}{2} H_2 + NaOH$$

Process tree

Level in process tree

Fig. 5.4 The chemical reaction and process tree for the creation of 1 kg of chlorine gas (the functional unit). The tree is separated into a number of levels (n) starting with the functional unit at $n = 0$ and expanding downward to include material and energy inputs required at each level. Note that the values presented here are arbitrary and only intended for illustration of the concept

According to the ISO 14044 standard, allocation should be approached in the order of:

1. Where it is possible, allocation may be avoided either by increasing the level of detail of the model or through *system expansion*. This can be done for multi-product systems. For example, Fig. 5.5 uses LCA to compare the incineration of 1 kg of waste solvent between two different incineration plants. Plants A and B, however, cannot be compared through the proposed functional unit (see figure) since Plant B also produces electricity as a coproduct. Through system expansion, the additional resources and emissions required to generate 0.33 kWh of electricity at Plant A can be either added to Plant A or subtracted from (credited to) Plant B.
2. Allocate resource uses and emissions among the coproducts according to physical/chemical relationships that exist between each coproduct and elementary flow. Such relationships can be based on, for example, the stoichiometry of a chemical reaction and described using *transfer coefficients*. For example, Fig. 5.6 shows the combustion of natural gas (x_1), propane (x_2), and air (x_3) to produce 1000 MJ of heat (y_1), which is the main product and the functional unit. However, the combustion also produces SO_2 (y_2), NO_x (y_3), and CO_2 (y_4). To allocate each of these four system outputs to the two inputs of natural gas and propane (air is not considered), the relevant transfer coefficients as shown in Fig. 5.6 can be used. For example, 55 kg of the total 59 kg of CO_2 produced by the combustion can be allocated to natural gas, with the remaining 4 kg allocated to the propane. Note that the transfer coefficients are based directly on existing

physical or chemical relationships that exist: (i) for heat produced on the specific heat of combustion of the inputs, (ii) for SO_2 on the sulfur content of the inputs, (iii) for NO_x both on the nitrogen content of the inputs and on the thermal NO_x produced that is dependent on the combustion temperature, and (iv) for CO_2 on the stoichiometry of the combustion reaction.
3. If no physical/chemical relationships exist, allocate according to other relationships, e.g., economic value or mass of the products.

If there are several options for allocation available, it is important to carefully compare the different options and transparently document the procedure taken.

Proposed functional unit: Disposal of 1 kg of waste solvent by incineration

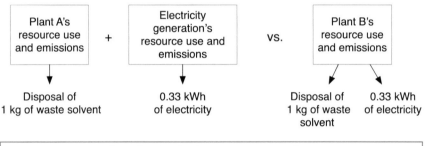

Functional unit after system expansion:
Disposal of 1 kg waste solvent plus the production of 0.33 kWh of electricity

Fig. 5.5 Example of applying system expansion to avoid allocation in an LCA comparing two incineration plants for the disposal of 1 kg of waste solvent

5.4.3 Data Quality

When collecting life cycle inventory data, it is important to obtain system-specific data that have the best possible quality. Table 5.1 summarizes some of the key data descriptors and examples of quality levels that can be associated with them. These include reliability, completeness, and temporal, spatial, and technical correlation with the system investigated. Completing mass balances of the system as well as comparing different data sources helps ensure that correct and complete data sets are used and a consistent level of quality can be maintained when building the inventory.

The inventory data for many activities (in particular for the production of basic materials and products, standard energy sources, transport, and disposal services) have already been collected and are included in large inventory databases available in either open-source formats (such as the US Life Cycle Inventory Database (NREL, 2012)) or subscription-based formats (such as Ecoinvent (Wernet et al., 2016)). A global map of inventory data sets available by country has also been put together (UNEP, 2019b) based on the openLCA Nexus online repository for LCA data (GreenDelta GmbH, 2019).

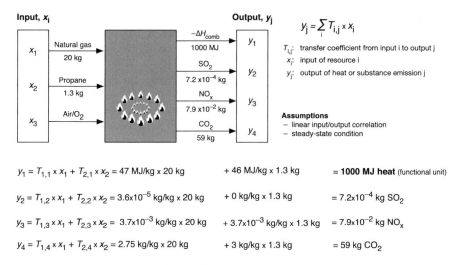

$$y_j = \sum_i T_{i,j} \times x_i$$

$T_{i,j}$: transfer coefficient from input i to output j
x_i: input of resource i
y_j: output of heat or substance emission j

Assumptions
– linear input/output correlation
– steady-state condition

$y_1 = T_{1,1} \times x_1 + T_{2,1} \times x_2 = 47$ MJ/kg x 20 kg $+ 46$ MJ/kg x 1.3 kg $= \textbf{1000 MJ heat}$ (functional unit)

$y_2 = T_{1,2} \times x_1 + T_{2,2} \times x_2 = 3.6 \times 10^{-5}$ kg/kg x 20 kg $+ 0$ kg/kg x 1.3 kg $= 7.2 \times 10^{-4}$ kg SO_2

$y_3 = T_{1,3} \times x_1 + T_{2,3} \times x_2 = 3.7 \times 10^{-3}$ kg/kg x 20 kg $+ 3.7 \times 10^{-3}$ kg/kg x 1.3 kg $= 7.9 \times 10^{-2}$ kg NO_x

$y_4 = T_{1,4} \times x_1 + T_{2,4} \times x_2 = 2.75$ kg/kg x 20 kg $+ 3$ kg/kg x 1.3 kg $= 59$ kg CO_2

Fig. 5.6 Example of using transfer coefficients to support allocation by describing the physical/chemical relationships between coproducts and elementary flows during combustion in a gas stove

Table 5.1 Some key data quality descriptors and examples of decreasing data quality from left to right (Weidema and Wesnæs, 1996)

Reliability	Verified data	Data on the basis of assumptions	Estimated data
Completeness	Representative data	Representative data for shorter periods	Fragmented data
Temporal correlation	Data less than 3 years from the year studied	Data less than 10 years from the year studied	Unknown or greater than 10 years from the year studied
Spatial correlation	Data from the location investigated	Data from a similar location	Data from a different or unknown location
Technical correlation	Data from investigated process/substance	Data from similar process/substance	Data from different or unknown process/substance

LCI data can often be obtained as average values or as local-, time- or technology-specific values, and many LCI databases are already integrated into popular LCA software packages such as SimaPro and OpenLCA. In contrast, company-specific data can often be obtained directly from an in-house process, product, or environmental information system. In compiling LCI data, it is important to pay attention to the temporal, geographic, and technological scope. If not enough information is available, it may be possible to:

- Use known yields and stoichiometry from an analogous process.
- Use thermodynamic estimates or internal company statistics for missing energy flows.

- Calculate material balances for missing waste flows and emissions; calculations can also be done to define missing degradation rates, transfer coefficients, etc. in treatment technologies.

5.4.4 Calculation Methods

Once all of the necessary data have been collected for the processes being analyzed (on the unit process level), the cumulative resource uses and emissions across the entire life cycle of the product or process under review can be calculated. The *sequential method* and the *matrix inversion method* are two common calculation approaches used to do this. They are each introduced here following the same example of chlorine production described previously in Fig. 5.4.

5.4.4.1 Sequential Method

The *sequential method* is a successive summation of the whole process chain starting from the foreground processes ($n = 1$) toward the background processes ($n = 2$, $n = 3$, etc.). Figure 5.7 shows how the flows to be considered expand toward the background processes. Production of the salt needed on the $n = 1$ level requires the use of rock salt, steam, steel, and potentially many other resources, and it also results in emissions on the $n = 2$ level. The steam needed on the $n = 2$ level requires its own set of resources and results in emissions on the $n = 3$ level.

In this way, the process tree can expand toward infinity with *recursive processes* that include material and energy resources needed for the construction and operation of all the infrastructure used. Recursive processes in this context are, for example, including the steel and electricity needed to construct a production plant, as well as the steel and electricity needed to manufacture each of the tools used during the plant's construction. In the end, all of these elementary flows from across the tree need to be summed together to produce an inventory of the total amounts of resources and emissions required for the functional unit (production of 1 kg of chlorine gas).

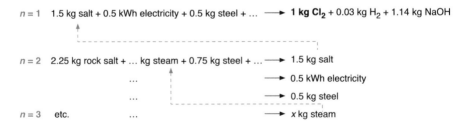

Fig. 5.7 Illustration of the expansion toward background processes when the sequential method is used to calculate the LCI for 1 kg of produced chlorine gas (the functional unit). All of the energy and resource uses across all levels and unit processes need to be cumulated

Depending on the set system boundaries, this method can clearly reach its practical limits quickly. To handle this, a calculation method using a linear system of equations in the form of matrices has been developed.

5.4.4.2 Matrix Inversion Method

The *matrix inversion method* uses a system of linear equations to calculate the cumulative resource uses and emissions of a system. This is done by setting up what is known as a technosphere matrix (**A**) and a biosphere matrix (**B**). Figure 5.8 illustrates the structure of both.

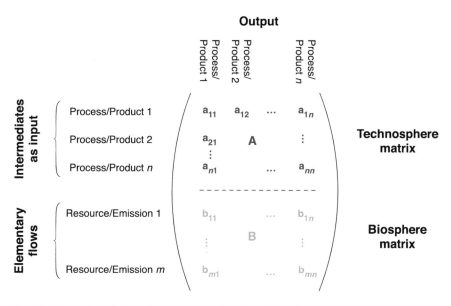

Fig. 5.8 Illustration of the technosphere matrix (**A**) and biosphere matrix (**B**) used within the matrix inversion method for developing the life cycle inventory (LCI)

The values in the technosphere matrix (**A**) represent the amounts of materials or resources (input) required for each process or product in the system (output). For example, the value of a_{12} in the matrix represents the amount of process or product 1 (in the row) needed as input to produce process or product 2 as an output (in the column). For the application of the matrix inversion method to the chlorine gas example, Fig. 5.9 shows how the values from the process tree are input into the technosphere matrix. For example, 1.5 kg of salt is required as an intermediate input per kg of chlorine gas produced.

The values in the matrix represent the first level of the process tree. When squared ($\mathbf{A}^2$), it represents the second level; when cubed ($\mathbf{A}^3$), it represents the third level; and so on ($\mathbf{A}^n$). As n approaches infinity, $\mathbf{A}^n$ goes to zero. To consider all of the different levels, the sum of all levels can be described mathematically as a Neumann series with the identity matrix (**I**) as $(\mathbf{I} - \mathbf{A})^{-1}$. The technosphere matrix can then

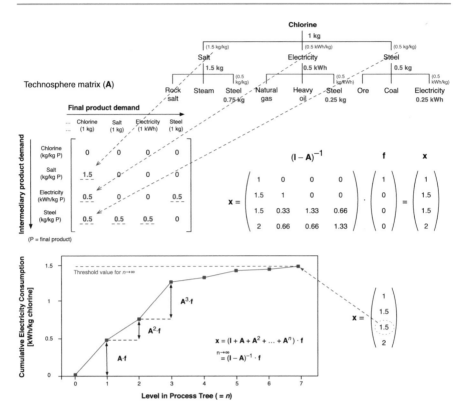

Fig. 5.9 Setup of the technosphere matrix (**A**) and calculation of the total product demand vector (**x**) using the technosphere matrix (**A**), the identity matrix (**I**), and the final product demand vector (**f**) for the production of 1 kg of chlorine gas

be used to calculate the total, cumulative demand of all products. This is known as the total product demand and is represented by the vector **x**. To calculate **x**, the Neumann series using the identity matrix needs to be multiplied by the final product demand vector (**f**), which describes the functional unit:

$$\mathbf{x} = (\mathbf{I} - \mathbf{A})^{-1} \times \mathbf{f} \qquad (5.1)$$

This mathematical approach allows for inclusion of all of the recursive processes across the different levels, which would have been tedious (and simply overwhelming) to include by using the sequential method introduced in the previous section. The inclusion of recursive processes is apparent by noticing that two of the values (1.33) in the diagonal of the matrix $(\mathbf{I} - \mathbf{A})^{-1}$ in Fig. 5.9 are greater than 1.

Figure 5.9 shows the final product demand vector (**f**) representing the 1 kg of chlorine gas produced, the resulting total product demand vector (**x**), as well as a plot illustrating the accumulation of electricity consumption across the levels of the process tree to reach the total of 1.5 kWh.

The values in the biosphere matrix (**B**) represent the elementary flows of energy or resources extracted from or emitted into the environment for each process or product in the technosphere matrix (**A**).

Figure 5.10 shows how the values from the process tree are input into the biosphere matrix. For example, 1.5 kg of rock salt is extracted as a resource from the environment to produce 1 kg of salt.

Combining the flows within the technosphere with the flows within the biosphere, these two matrices can be used together with the final product demand vector (**f**) to calculate the total elementary flows from and to the environment (**y**) across all levels of the process tree:

$$\mathbf{y} = \mathbf{B} \times (\mathbf{I} - \mathbf{A})^{-1} \times \mathbf{f} \qquad (5.2)$$

Figure 5.10 shows how the elementary flow values from the process tree are input into the biosphere matrix and used to calculate the total elementary flow vector (**y**) for the production of 1 kg of chlorine gas.

Following the same approach as introduced for the technosphere matrix, Fig. 5.10 illustrates how the cumulative emissions of 2.75 kg CO_2 are reached when totaling along all of the levels within the process tree.

5.4.5 Direct Interpretation of LCI Results

The developed life cycle inventory provides information on the use of materials and resources (including energy) of the system under investigation, and sometimes it provides enough information to already highlight some key areas for improvement. For example, comparing the energy and resource use of a chemical product or process from the inventory with theoretical optima of the reaction can provide insight into:

- Energy-saving potential
- Increasing synthesis yield
- Better use of input resources
- Reducing refinement and application losses
- Current capacity bottlenecks

LCI data can sometimes also already be used to compare alternatives. For example, Fig. 5.11 shows a set of alternative reducing agents that can be used for the reaction from a monoazo compound to TINUVIN®. Through the development of an LCI, the by-products from the reduction reaction can be determined to range from clearly more hazardous substances such as zinc hydroxide ($Zn(OH)_2$) to benign substances such as water. Of the three produced by-products, water is the least reactive and also has the lowest molecular weight. This means that with water, there will also be a lower quantity of by-product to manage. Considering this, H_2 might be the better reducing agent to use in this application. However, to achieve a complete

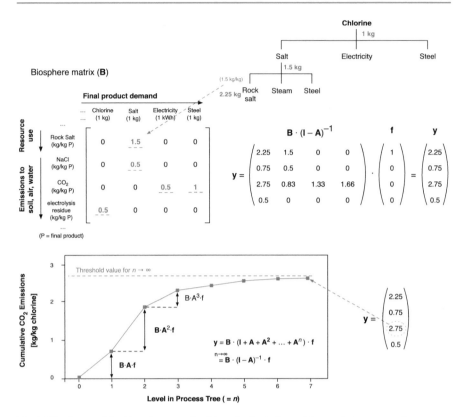

Fig. 5.10 Setup of the biosphere matrix (**B**) and calculation of the total elementary flow vector (**y**) using the biosphere matrix (**B**), the technosphere matrix (**A**), the identity matrix (**I**), and the final product demand vector (**f**) for the production of 1 kg of chlorine gas

life cycle perspective before making a decision, additional aspects such as the energy required for the production of the agents should also be compared.

While the direct interpretation of LCI results can sometimes be helpful and point toward improvements, the next phase of LCA is still needed to calculate the associated health and environmental impacts caused.

5.5   Impact Assessment (Phase 3)

In the third LCA phase of *life cycle impact assessment* (LCIA), the cumulative amounts of material and energy calculated during the previous LCI phase are placed within *impact categories*, and their impacts are calculated. Through this process, the diverse and often copious information from the LCI is translated into just a few *indicators*. Often, these indicators are relatively simple to identify and can communicate representative information to effectively illustrate health and

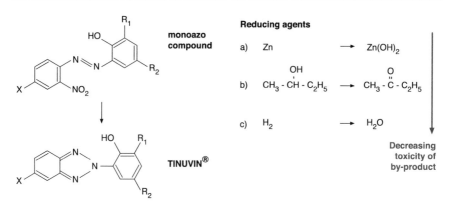

Fig. 5.11 Example of using LCI results to select a best reducing agent through identifying and comparing their by-products in the reaction from a monoazo compound to a TINUVIN® light-stabilizing compound

environmental impacts. Optional further steps include to *normalize, group*, and/or *weight* these indicators, making it possible to achieve a more refined set of results for simpler comparison of previously complex and diverse information.

The impact assessment phase can be divided into four main steps as outlined in Fig. 5.12.

Fig. 5.12 The main steps of a life cycle impact assessment (LCIA)

A preparatory first step is the (i) *selection* of the impact categories, indicators, and characterization models that will be used to define them. These should be chosen based on the defined goal and scope set for the LCA. (ii) In the next *classification* step, the flows calculated from the LCI are assigned to the chosen impact categories. (iii) In the *characterization* step, model-based factors are then used to convert the assigned flows into their category-specific indicators. This means that the LCI is in a linear relationship with the corresponding life cycle impacts.

Following the ISO standard, these results can then optionally undergo (iv) *normalization* to adjust them onto a scale representative of a specific region or population and/or *grouping* and *weighting* to aggregate multiple indicators into fewer results. The following subsections introduce each of these steps within LCIA, and further explanations and examples can be found in Hauschild and Huijbregts (2015); Verones et al. (2017), and UNEP (2016).

5.5.1 Step 1: Selection of Impact Categories, Indicators, and Characterization Models

In this first step, impact categories need to be selected that are of interest and help to answer questions set within the goal of the LCA. For example, if the LCA aims to optimize the natural resources (both abiotic and biotic) used during a production process, then impacts related to, for example, water use, fossil resource use, and land use might be good impact categories to select. These would all be examples of *midpoint* impact categories, which are representative of intermediate impacts along the pathway to calculating *endpoint* damages to areas of protection such as human health, ecosystems, and resource availability.

In this chapter, the ReCiPe impact assessment method (Huijbregts et al., 2017) is often used to help introduce this phase of LCA, and Fig. 5.13 illustrates the midpoint and endpoint categories considered within this method.

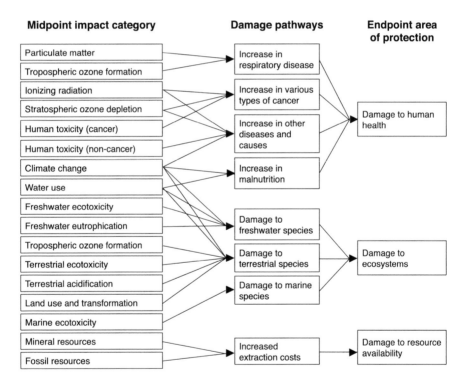

Fig. 5.13 Schematic representation of the midpoint impact categories as well as the connected damage pathways and resulting endpoints within the ReCiPe impact assessment method (Huijbregts et al., 2017)

ReCiPe is just one of the many different methods developed for assessing impacts. Each method contains calculated *characterization factors (CF)* that are used to convert material or energy flows into midpoint or endpoint impacts. While these are very helpful to use, CFs can differ significantly from one another in how they characterize the flows, the impact models they use, and how they might optionally normalize, group, or weight different impacts.

In selecting impact categories, indicators, and methods to use for an LCA, it is helpful to critically compare and consider available impact assessment methods using the following basic criteria:

- Completeness: Are all conceivable effects on health and the environment taken into account by the method?
- Practicality: How easy, time-efficient, and cost-efficient is the method? Is any aggregation of the individual impacts reproducible?
- Scientific content: Is the latest science in terms of effects and mechanisms of action considered? Are the local and temporal effects as well as the different routes of exposure appropriately taken into account? Finally, is there a clearly defined relationship between the various effects and potential damage?
- Transparency: Does the method evaluate the impacts in a traceable way?

The Life Cycle Initiative led by the United Nations Environment Programme (UNEP) and the Society of Environmental Toxicology and Chemistry (SETAC) has also published guidance documents on the use of impact assessment indicators in LCIA (UNEP, 2016, 2019a).

Another relevant tool within impact assessment and used in this chapter is the USEtox model (Rosenbaum et al., 2008; Hauschild et al., 2008). Developed with the support of the UNEP-SETAC Life Cycle Initiative (UNEP, 2020), USEtox is an environmental model that estimates the fate, exposure, and effects of chemicals to characterize their human health and ecotoxicological impacts. It is the result of an effort to gain scientific consensus and harmonize the multiple characterization models for chemicals that existed in the past. Currently, it is being implemented both on its own and within various LCA impact assessment methods.

Some of the other impact assessment methods that are widely used today include LC-IMPACT (Verones et al., 2019), IMPACT World+ (Bulle et al., 2019), Ecological Scarcity (Frischknecht and Büsser Knöpfel, 2013), and TRACI (Bare, 2011).

5.5.2   Step 2: Classification

In the *classification* step, the flows from the life cycle inventory are assigned to the chosen midpoint impact categories. Some common, chemically related midpoint impact categories include:

- Climate change
- Trophospheric ozone creation
- Freshwater eutrophication

- Human toxicity

It is important to note that a single flow can potentially contribute to more than one of the impact categories chosen.

5.5.3   Step 3: Characterization

In the *characterization* step, elementary material and energy flows are multiplied by a *characterization factor* for each of their classified impact categories. This results in the calculation of impacts and allows the different flows to be compared to one another with a common unit. Scientifically based characterization factors are used within each available impact assessment method. They are often calculated using an effect-based comparison with a reference substance.

The calculation of the characterization factors is different for each impact category (and sometimes also among different impact assessment methods), and a dose-effect relationship is used to define each. For a midpoint impact category i, the impact (I_i) can be calculated across all material and energy flows (f_j) by using the corresponding characterization factors ($CF_{i,j}$) as:

$$I_i = \sum_j CF_{i,j} \times f_j \tag{5.3}$$

- I_i: impact within impact category i
- f_j: material or energy flow j (from LCI)
- $CF_{i,j}$: characterization factor (impact assessment method specific) for the conversion of material or energy flow f_j into impact equivalents of the impact category i

For the midpoint impact category of climate change, for example, Eq. 5.3 would be applied with the characterization factor for *global warming potential* (GWP) (see Eq. 5.7 for details):

- $I_{\text{climate change}} = \sum_j CF_{\text{GWP},j} \times f_j$

Characterization factors can also be derived to translate impacts from midpoints to endpoints. Within the ReCiPe method, for example, midpoints are set to contribute to three endpoint areas of protection: human health (expressed in units of disability-adjusted life years (DALY)), ecosystems (expressed in units of biodiversity loss), and resource availability (no harmonized unit set). The method provides users with midpoint-to-endpoint characterization factors that can be used in the same way to calculate damages to these three endpoints.

The following sections provide examples that briefly illustrate how such characterization factors can be developed and applied.

5.5.3.1 Cumulative Energy Demand

In LCA, the *cumulative energy demand (CED)* of a product or process is the total amount of both direct and indirect uses of *primary energy* involved in the extraction, manufacturing, and disposal of all raw and auxiliary materials. Primary energy includes energy found in the natural environment before being converted into any other form (e.g., crude oil, natural gas, solar energy, wind energy). Because the energy use within chemical manufacturing can be high, CED is an important aspect to consider. While there have been ongoing discussions to agree upon an appropriate approach to calculating and applying CED within LCA (Frischknecht et al., 2015; Huijbregts et al., 2010, 2006), it is considered to fit within LCIA, rather than within the LCI phase.

Individual characterization factors can be calculated to define the CED of manufactured products and of process steps. One study modeled and calculated the CED of a range of chemical solvents commonly used in industrial processes (Capello et al., 2007). Applying a functional unit of 1 kg of solvent used as reaction media in a process, the authors calculated the CED for the production of each virgin solvent, incineration of each solvent as waste, and distillation of each solvent if it were recycled. Credits were given for the production of energy from the incineration and also for avoiding the production of virgin solvent through recycling.

Figure 5.14 diagrams an example of how the CED of a process system can be impacted through recycling the solvent used. Here, it is assumed that (i) no change of chemical properties occur from primary to secondary solvent, (ii) no solvent loss occurs along the use and distillation steps, and (iii) there is a constant recycling rate (r) with a value between 0 and 1.

Based on the energy required for primary solvent production, the distillation of the solvent for reuse, and waste solvent incineration, the CED for this system can be determined in relationship to the recycling rate (r).

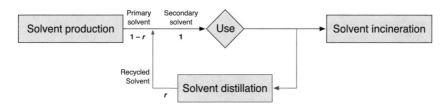

Fig. 5.14 Calculation of the cumulative energy demand (CED) for a chemical process depending on the fraction of solvent recycled. The recycling rate (r) can vary from 0 to 1

The CED for the system is represented by the CED of the secondary solvent ("sec.solv."). This can be derived from the CEDs of the primary solvent ("prim.solv."), solvent distillation ("solv.dist."), and solvent incineration ("solv.incin."):

$$CED_{\text{sec.solv.}} = (CED_{\text{prim.solv.}} + CED_{\text{solv.incin.}}) \times (1 - r) + CED_{\text{solv.dist.}} \times r$$
(5.4)

Generally, recycling through distillation is preferable to incineration when:

$$CED_{\text{prim.solv.}} + CED_{\text{solv.incin.}} > CED_{\text{solv.dist.}}$$
(5.5)

This application of solvent recycling is an example of a process that can support a transition toward a *circular economy*, which aims to design out waste and pollution, to keep products and materials in use, and to regenerate natural systems (Ellen Macarthur Foundation, 2019).

5.5.3.2 Climate Change

The global warming that causes climate change derives from the radiative efficiency of the greenhouse gases present in the atmosphere. Fundamentally, Earth's dynamic heat balance can be defined as shown in Eq. 5.6 by considering the absorbed incoming solar radiation and the outgoing radiation. These energy flows have been calculated on a global level by many studies, including Trenberth et al. (2009):

$$\underbrace{C\frac{dT}{dt}}_{\text{disruption to radiative equilibrium}} = \underbrace{\frac{S_0}{4}(1 - a(T))}_{\text{absorbed incoming solar energy}} - \underbrace{\epsilon(T)\sigma T^4}_{\text{outgoing radiative flux}}$$
(5.6)

- C: perturbation-related heat capacity of the Earth [J m^{-2} K^{-1}]
- T: transient temperature, here as response to the radiative forcing of greenhouse gases [K]
- S_0: solar radiation flux ($\frac{1}{4}$ is the correction for projected planet area (= πr^2) to planet surface area (= $4\pi r^2$)) [W/m^2] with r being the radius of the Earth
- $a(T)$: fraction of radiation that is immediately reflected back into space, known as albedo ($\cong 0.3$) [–]
- $\epsilon(T)$: long-wave emissivity ($0 \leq \epsilon \leq 1$) [–]
- σ: Stefan-Boltzmann constant (= 5.67×10^{-8}) [W m^{-2} K^{-4}]

Much of the incoming solar energy is received in the form of UV-VIS radiation, whereas the outgoing radiative flux is mostly in the form of infrared radiation. Higher concentrations of greenhouse gases in the atmosphere lead to lower values of the long-wave emissivity (ϵ in Eq. 5.6).

A substance's global warming potential (GWP) relative to the GWP of CO_2 serves as the characterization factor for the climate change impacts of this substance. This characterization factor can be quantified through Eq. 5.7 with CO_2 as the reference substance and with the characterization factor defined in units of CO_2

equivalents. Examples of the calculated 100-year GWP for various substances include CH_4 (28 kg CO_2-eq./kg), N_2O (265 kg CO_2-eq./kg), and CCl_2F_2 (10,200 kg CO_2-eq./kg) (Myhre et al., 2013):

$$CF_{GWP,j} = \frac{\int_0^T e_j \times c_j(t)\, dt}{\int_0^T e_{CO_2} \times c_{CO_2}(t)\, dt} \qquad (5.7)$$

- $CF_{GWP,j}$: global warming potential as the characterization factor of climate change for substance j [kg CO_2-eq./kg j]
- e_j: radiative efficiency change due to a unit increase in atmospheric concentration of substance j [W m^{-2} kg^{-1}]
- $c_j(t)$: time-dependent concentration of substance j after an instantaneous release [kg/m^3]
- T: time frame considered (often set as 100 years)

5.5.3.3 Photochemical Ozone Creation

Another midpoint impact category is the potential of an emitted substance to create ozone in the troposphere. Ozone is created from nitrogen oxides (NO_x), which are emitted from vehicles and other sources of combustion, and volatile organic compounds (VOCs), which are emitted from incomplete combustion, industrial processes, and vegetation. In the presence of sunlight, these chemicals can react with one another and produce ground-level ozone, i.e., in the troposphere, which can be damaging to human and environmental health. The amount of ozone generated depends on the concentration of NO_x and VOCs in the atmosphere, as well as on the meteorological conditions.

The characterization factor for *photochemical ozone creation* is based on the reference substance ethylene (C_2H_4). Figure 5.15 illustrates the generation of ozone (O_3) through reaction of VOCs (depicted as RCH_3 in the figure), NO_x, and hydroxyl radicals ($\cdot OH$). The primary source of tropospheric hydroxyl radicals in the environment is a pair of reactions starting with the photodissociation of ozone by solar UV radiation. Equation 5.8 defines the characterization factor for the photochemical ozone creation potential (POCP) using ethylene as the reference substance:

$$CF_{POCP,j} = \frac{\delta[O_3]_j}{\delta[O_3]_{Ethylene}} \qquad (5.8)$$

- $CF_{POCP,j}$: characterization factor representing tropospheric photochemical ozone creation potential (POCP) of substance j [kg C_2H_4-eq./kg j]
- $\delta[O_3]_j$: tropospheric ozone concentration change per kg of substance j emitted into the atmosphere (integrated over a time period of 0 to 5 d)

Within the ReCiPe method, POCP characterization factors that were developed under European conditions with a 5-day trajectory model for various products

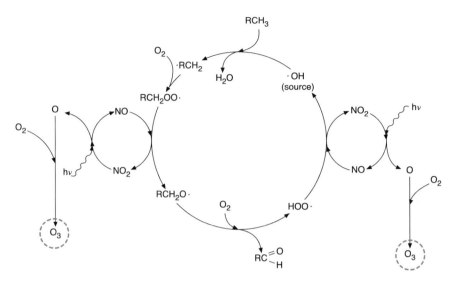

Fig. 5.15 Illustration of the mechanism for tropospheric photochemical ozone (O_3) creation from a reaction of hydroxyl radicals with volatile organic carbons (VOCs, shown as RCH_3) and nitrogen oxides (NO_x). $h\nu$ denotes the energy of a photon of light of the frequency ν. h is the Planck quantum of action (in J s)

and processes (Derwent et al., 2007) were used to define midpoint and endpoint characterization factors describing impacts on human health from generated ozone (RIVM, 2017).

5.5.4 Step 4: Optional Normalization, Grouping, and Weighting

Important conclusions can often be drawn just by looking at the results from the midpoint or endpoint indicators as they are. However, the optional impact assessment steps of *normalization*, *grouping*, and *weighting* can help to further refine the LCA results and make them more comprehensible or easier to compare.

5.5.4.1 Normalization
Through *normalization*, the results of an impact at the midpoint or the endpoint are calculated to be relative to a selected reference. Doing this can help to show the relative magnitude of an impact in comparison to a total impact for a region or population, show different impacts on a common scale, and help to check that the calculated impacts are reasonable (and not the result of an error in the calculations). Normalization is often carried out, for example, with reference to:

- Total impacts
- The population, e.g., to calculate impacts per person

- A reference state, e.g., natural background pollution, current environmental impact for an area, etc.
- Critical values, e.g., legal limits, scientifically defined acceptable limits, etc.
- A reference product, reference process, etc.

The impact from an individual impact category j (I_j) is therefore set in relation to the impact of that category for the reference system (I_{ref}). This yields the normalized impact (NI_j):

$$NI_j = \frac{I_j}{I_{ref}} \tag{5.9}$$

Within the ReCiPe impact assessment method (Huijbregts et al., 2017), the global annual impacts in the year 2000 are available as reference systems for normalization. The total calculated impacts that took place globally in that year for each impact category were calculated and used to define the corresponding normalization factor I_{ref} (Sleeswijk et al., 2008). Examples of some of these developed factors are shown in Table 5.2. Normalization references have also been developed specifically for use with characterization factors within the USEtox model (Laurent et al., 2011).

Table 5.2 Impact categories and their corresponding reference substances and global normalization factors based on the year 2000 (Sleeswijk et al., 2008). PM_{10} is the amount of particulate matter in air less than $10\,\mu m$ in diameter. DCB is 1,4-dichlorobenzene

Impact category	Reference substance (equivalents)	Normalization factor (I_{ref})
Climate change (100 yr)	kg CO_2 eq.	4.18×10^{13}
Freshwater eutrophication	kg phosphorus	3.77×10^{9}
Human toxicity (100 yr)	kg 1,4-DCB	1.20×10^{12}
Terrestrial ecotoxicity (infinite time)	kg 1,4-DCB	5.09×10^{10}
Particulate matter formation	kg PM_{10}	9.92×10^{10}

5.5.4.2 Grouping

A simple approach to analyzing the impact assessment results can be through sorting or ranking them. This is called *grouping* and can be done in various ways. For example, some impact categories included in an LCA could be grouped based on the type of resources or emissions involved, by a particular protected good, or by spatial scale (e.g., local vs. global impacts). A defined hierarchy based on societal values can also be used to prioritize the impact categories during interpretation.

5.5.4.3 Weighting

For LCA results to be more directly comparable with one another, the midpoint or endpoint impacts can be *weighted* to aggregate them and achieve a single score. This requires comparing all considered impact categories and deciding which are most (and least) important. Experts can be consulted to help do this, and setting political targets or societal priorities can also be used.

The impacts (I) of individual midpoints or endpoints (j) can be weighted by multiplying them by set weights (w) and then summing them to produce a single total impact score (I_{total}):

$$I_{total} = \sum_j I_j \times w_j \qquad (5.10)$$

While the selection of the impact categories already involved a certain level of subjective user input, the weighting of the impact categories is often seen as unscientific since it is subjective and based on personal and societal perception patterns and value systems that can vary according to place and time.

The validity of a weighted life cycle assessment is therefore limited to the people who agree with the weightings used. According to the ISO 14044 standard, fully aggregated methods involving weighting may be used in LCA for product optimization, but they are not allowed to be used for results presented publicly for product comparison. It is also important to remember that if one product or process alternative being compared has a better score in all midpoint or endpoint impact categories compared to the other alternatives, then weighting is logically not necessary. Current approaches and ongoing discussions regarding the application of normalization and weighting within the LCA community are presented in more details by Hauschild and Huijbregts (2015) and Pizzol et al. (2017).

When it seems necessary to aggregate results before a decision can be made, *multi-criteria decision analysis (MCDA)* is an additional method that can help. MCDA is a framework that supports making complex decisions with multiple and often conflicting objectives that are valued differently by stakeholder groups involved. The framework was developed in the field of operations research, and numerous studies and guidelines exist that describe the specific approaches available. For an overview of MCDA, see Hermans and Erickson (2007) and Zanghelini et al. (2018).

5.6   Interpretation (Phase 4)

The last of the four phases of LCA focuses on the review and *interpretation* of the results from each of the previous three phases. While a thorough review should be completed iteratively after completion of each LCA phase, the overall aims of this phase at the end of the LCA are to:

- Identify the most important environmental flows, such as key process parameters, resource consumption or emissions, significant impacts, etc.
- Evaluate the completeness, sensitivity, and consistency of the study
- Develop final conclusions and recommendations

In doing so, it is also valuable to consider and comment on the influence that the chosen allocation and impact assessment methods may have had on the results, as well as on the set system boundaries. In the context of integrated product or process development, it is especially useful to identify which process steps within the entire process tree are the dominant contributors to the total impacts. Knowing this can provide clues for how to improve the design and reduce the impact as much and as efficiently as possible.

This phase is also a good time to systematically check the data, assumptions, and calculations used within the LCA for errors, proper structure, and good documentation. Look to ensure that the units used are consistent and mass and energy balances within the considered system are correct. It is also important to complete a *sensitivity analysis* to identify the most influential model variables and assumptions within the LCA and their impact on the results. Various methods and approaches exist to analyze sensitivity, each with different levels of complexity and detail of insight provided in their results. An introduction to and discussion of some of the most commonly used methods within LCA are available in Jolliet et al. (2015).

5.7   Example: Comparing Insulation Foaming Agents

With the four phases of LCA introduced, this section provides a practical example of completing a simple LCA to compare a pair of alternative foaming agents for use in extruded polystyrene (XPS) insulation in buildings. Each of the four phases within LCA is described, and base data are provided for the necessary calculations.

5.7.1   Goal and Scope Definition (Phase 1)

The *goal* of the LCA in this example is to identify the best foaming agent for use in XPS to insulate a building. In particular, this will be a comparison of existing foaming agent alternatives (and not a prospective LCA considering future technologies). The target audience of the LCA results is XPS insulation manufacturers to help them select among alternative foaming agents during production. Figure 5.16 shows the entire system for the foaming agents including production, use, and recycling/disposal. The specific *scope* considered within this LCA is highlighted in the figure and only includes resource use and emissions from the (i) production and (ii) use of the foaming agent. The system elements of polystyrene production, foaming of the polystyrene, and recycling/disposal are considered to be the same regardless of the foaming agent used, and they are therefore not included in this LCA.

The *functional unit* of the LCA is defined to be XPS insulation sufficient for the foundation of a building with a base of $100\,m^2$ and an expected lifetime of 50 years. Several of the previously introduced methods and characterization factors will be applied, and life cycle inventory and impact data sets hosted by the Ecoinvent v3.5 database (Wernet et al., 2016) will be used to calculate the impacts within the midpoint categories of climate change, cumulative energy demand (CED), and ecotoxicity.

Fig. 5.16 Schematic of the entire system with only the production and use of the foaming agent highlighted in gray as being within the scope of this LCA

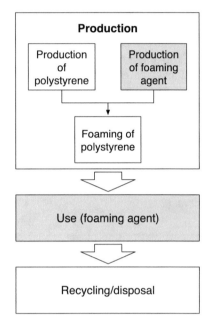

5.7.2   Inventory Analysis (Phase 2)

Table 5.3 shows information that was found to be available to describe the XPS insulation itself and the two alternative foaming agents that can be used to produce it: pentane and CO_2.

Table 5.4 shows the resulting data found upon completing the inventory analysis for each of the alternative foaming agents. Part A shows the calculated amount of each foaming agent required for the production of the XPS insulation to achieve the functional unit, and part B shows the calculated amount of each foaming agent released over the expected 50-year lifetime of the XPS insulation once installed.

In Table 5.4, the amounts of foaming agent emitted over the course of the use phase of the XPS insulation were calculated using the expected lifespan of XPS insulation and the half-lives for each foaming agent as given in Table 5.3.

Table 5.3 Base data for the life cycle inventory analysis regarding the two alternative foaming agents

Insulation	
XPS insulation required for a building with an area of 100 m^2	50 m^3
Expected lifetime of the insulation	50 years
Foaming agent	
Amount of foaming agent required	
Pentane	4 kg/m^3
CO_2	2 kg/m^3
Half-life ($t_{1/2}$) of foaming agents within insulation	
Pentane	50 years
CO_2	<1 year

Table 5.4 Life cycle inventory: (A) Amount of each foaming agent required per functional unit insulation and (B) calculated emissions of each foaming agent from insulation over its expected lifetime of 50 years

	Pentane	CO_2
(A) Production		
Amount produced per functional unit insulation [kg]	200	100
(B) Use		
Emission during use [kg]	100	100

5.7.3   Impact Assessment (Phase 3)

With the inventory values collected and calculated, they can now be assigned to impact categories (*classification*), translated into midpoint impacts (*characterization*), and optionally further aggregated with the use of *normalization* and *weighting*.

Table 5.5 shows the midpoint characterization factors for climate change using the IPCC 2013 100-year method, total cumulative energy demand (CED), and ecotoxicity from the USEtox model (Henderson et al., 2011).

The CFs used to calculate impacts from the production of each of the foaming agents were all extracted from the Ecoinvent database. GWP factors for emitted gases into the atmosphere during the use phase are available in the Intergovernmental Panel on Climate Change's Fifth Assessment Report (Myhre et al., 2013), and ecotoxicity characterization factors for emitted gases are pre-calculated in the USEtox v2.12 model.

The CFs applied here from USEtox are expressed in comparative toxic units (CTU). For ecotoxicity (CTU$_e$), this is defined as aquatic ecotoxicity based on the estimated potentially affected fraction of species (PAF) from the emission integrated over time and volume [PAF $\times$ m^3 $\times$ day/kg chemical emitted]. In calculating these factors in the USEtox model, emissions of the foaming agents during the use phase were assumed to be released into urban air. The characterization factor for the ecotoxicity of CO_2 is set to be 0.

Because CED is the cumulative energy demand to produce the foaming agents and no additional energy is involved in the emission from the insulation during the use phase, it is set to be 0.

Table 5.5 Values of the characterization factors used to calculate climate change (GWP), total cumulative energy demand (CED), and aquatic ecotoxicity midpoint impacts for the alternative foaming agents. CED is in units of megajoule equivalents [MJ-eq], and ecotoxicity is in comparative toxic units [CTU]. Characterization factors for production are defined under the Ecoinvent database activity names "pentane production" by Hischier (2007) and "carbon dioxide production, liquid" by Althaus et al. (2007) for Europe using the cutoff by classification system model. (A) per kg produced, (B) per kg emitted

	CF_{GWP} [kg CO_2 eq./kg]	CF_{CED} [MJ-eq/kg]	$CF_{ecotoxicity}$ [CTU_e/kg]
(A) Production			
Pentane	1.1	86.6	0.2
CO_2	0.8	9.6	3.4
(B) Use			
Pentane	ca. 5	0	2.0×10^{-2}
CO_2	1	0	0

By multiplying the inventory amounts in Table 5.4 with the characterization factors in Table 5.5, the midpoint impacts can be calculated as shown in Table 5.6.

Table 5.6 Calculated midpoint impacts from the production phase and use phase, as well as the combined total impact (production + use) for each midpoint impact and alternative foaming agent

	Climate change [kg CO_2 eq.]	CED [MJ eq.]	Ecotoxicity [CTU_e]
(A) Production			
Pentane	220	1.73×10^4	40
CO_2	80	960	340
(B) Use			
Pentane	500	0	2
CO_2	100	0	0
(C) Total			
Pentane	720	1.73×10^4	42
CO_2	180	960	340

These calculated midpoint impacts can be further translated into endpoint impacts, which could then be even further *weighted* to compare impacts between foaming agents using a single indicator. However, aggregation to endpoints and further weighting will not be applied in this example, and instead, the midpoint impacts will be used for interpretation of the results.

Once calculated, midpoint (or endpoint) impacts can then be (optionally) *normalized*. Table 5.7 shows the results of normalizing the climate change impacts using the normalization factor of 4.18×10^{13} kg CO_2 from Sleeswijk et al. (2008). This factor represents the total global greenhouse gas emissions over a 100-year time horizon (see Table 5.2). For pentane, for example, this can be calculated as:

$$\frac{720 \text{ kg } CO_2 \text{ eq.}}{4.18 \times 10^{13} \text{ kg } CO_2 \text{ eq.}} = 1.72 \times 10^{-11}$$

The resulting normalized value shows the very small (but quantifiable) contribution that the use of these foaming agents have in comparison to total global greenhouse gas emissions. It is important to remember that normalization is done through the use of a constant and, hence, there is no difference in the outcome of a study before and after normalization.

Table 5.7 Normalized climate change impacts calculated for each foaming agent

	Normalization factor (Global, 100 yr horizon) [kg CO_2 eq.]	Pentane [–]	CO_2 [–]
Climate change (production and use)	4.18×10^{13}	1.72×10^{-11}	4.31×10^{-12}

5.7.4 Interpretation (Phase 4)

With all of the midpoint impacts calculated, the LCA results can now be interpreted. Upon reviewing the results in Table 5.6, some key takeaway messages are apparent:

- CO_2 has lower impacts than pentane across the midpoints of climate change and CED, but not for ecotoxicity.
- From these results, a clear recommendation cannot be determined between pentane and CO_2.

Further analysis is needed to make a decision on which foaming agent is best to use. This could be done, for example, through (i) calculation of additional midpoint impacts for each foaming agent such as ozone formation and water use, (ii) aggregation to endpoints such as damage to human health and ecosystems, and/or (iii) applying weighting to midpoints or endpoints based on political targets. Some additional open questions that were not considered within the scope of this LCA could also be helpful to take into account:

- Is it possible to recycle and reuse XPS insulation boards at the end of their expected lifetime? Could this then offset the need for the production of virgin boards?

- Does either of the foaming agents provide additional value for the insulation (e.g., heat-insulating properties)?
- How do the costs compare among the production of the foaming agents and the processes needed to use them? Would these then have a justifiable influence on the selection?

5.8   Eco-efficiency

The first guiding principle of integrated development is the implementation of *eco-efficiency*, which is short for ecological efficiency. It is defined according to the ISO 14045 standard (ISO, 2012) and is the ratio between the value of a product or process and its environmental impact along its entire life cycle.

As introduced in the earlier parts of this chapter, the environmental impacts can be calculated through the application of an LCA. The value, however, can be assessed according to the ISO standard in terms of a provided function, monetary amount, or even as an intangible contribution such as cultural or historical value. In the context of this book, the focus will be placed on the monetary value generated, and a fundamental approach for doing this is the calculation of the *net present value*.

5.8.1   Calculating Net Present Value

Economic accounting aims to determine the monetary value added along the life cycle of a product or process using the market prices of the goods and services involved. For this, a product's value (expressed in terms of market volume and market price) over a certain period is offset against the costs involved in bringing it to the market (including both variable and fixed costs) (Brealey et al., 2017). This difference is known as the *net present value (NPV)* and helps to take into account the time value of money. It is the total present value of a time series of cash flows covering the whole life cycle of a functional unit. Profit-seeking enterprises always aim to maximize the NPV given the allocated amount of funds available for the project.

The NPV is often calculated to also take into consideration costs associated with potential future risks, economic inflation, and the opportunity cost lost by future money not being able to accrue interest from being invested. This devaluing of future revenues and expenses is called *discounting*, and it is necessary because having 1 dollar today is preferred and valued more than having 1 dollar in the future. The exact amount to discount a future cash flow is subjective and is defined through an applied *discount rate*. If, for example, the uncertainty of future revenues and expenses is high, this can be reflected by applying a higher discount rate.

The NPV can be calculated by summing the discounted cash flows over the entire life cycle as:[1]

$$NPV = c_0 + \sum_{n=1}^{N} \frac{c_n}{(1 + r_n)^n} \qquad (5.11)$$

- *NPV*: Net present value [\$/functional unit]
- c_0: Cash flow at the present time ($t = 0$); value is often negative representing cash outflow as an initial investment [\$/functional unit]
- c_n: Cash flow in future year n; period-specific difference between revenues and costs; value is often positive representing a cash inflow of operating profit [\$/functional unit]
- r_n: Discount rate in year n; dependent on the rate of return in the capital market and on the specific risk of the project [dimensionless]
- N: Expected lifetime of an investment (e.g., of a chemical product or process) [yr]

For example, the NPV of a one-time \$1 million cash flow 35 years in the future is valued at just \$35,600 today with a 10% discount rate and at \$500,000 today with a discount rate of 2%. This shows that the relevant time horizon of an investment becomes shorter and shorter as the discount rate increases.

5.8.2 Influence of Integrated Development on NPV

In the context of life cycle costing, implementing safety and environmental protection through integrated development can have significant and beneficial impacts on a project's net present value. Some of these can include:

- Improvement of energy and resource efficiency: There may be an increase in investment costs in the short term (i.e., c_0 in Eq. 5.11) for more energy- and resource-efficient equipment and processes, but this tends to result in an increase of future cash flow (c_n) through long-term cost savings.
- Increase in inherent product and process safety: Lower-risk levels for a product or process mean that a lower annual discount rate can be applied (r_n) and an increase in NPV is obtained.

---

[1] Other common indicators for evaluating an investment include:

(i) *Payback time*: the time it takes for the NPV to become 0

(ii) *Internal rate of return (IRR)*: the discount rate that causes the NPV to become 0

- Gain in social acceptance: Through successfully increasing social acceptance of a product or process, faster regulatory approval procedures can often be expected, meaning less time until the product is on the market and until the cash inflow can become positive.

However, when not properly applied, integrated development can potentially have direct, negative impacts on the NPV such as:

- Maximization of solely safety, human health, and environmental protection: In this case, financial trade-offs are not considered carefully and will result in losing synergies with the business side of the project, higher total costs, and a negative impact on the NPV.
- Safety and environmental protection addressed after development: If integrated development is not followed from early stages of the design process, this can lead to additional costs that reduce the NPV, including (i) loss of valuable resources resulting from increased waste and emissions, (ii) additional investment costs for extra disposal facilities that could have been avoided, and (iii) additional operating and maintenance costs that could have been avoided.

As with LCA and any other type of assessment, it is important to also analyze and consider the sensitivity of the variables and inputs involved. In the case of calculating NPV, a sensitivity analysis can be performed to account for the uncertainty of expected costs and to compare actual costs with theoretical optimum values. For example, the effect of changing the product price or sales volume (by $\pm$ $X\%$) will lead to a specific change in the NPV (by $\pm$ $Y\%$).

5.8.3 Calculating Eco-Efficiency

If a chemical product or process has been evaluated both for its value added (e.g., through calculating the NPV) and for the life cycle impacts it has on human health and the environment (through completion of an LCA), these two quantities can be related to each other to determine the eco-efficiency (EE):

$$EE = \frac{\text{Net Present Value (NPV)}}{\text{Life Cycle Impact}} = \frac{NPV}{I} \qquad (5.12)$$

- EE: Eco-efficiency [$/impact]
- NPV: Net present value [$]
- I: The calculated human health and/or environmental impact [units depend on impact category]

To be defined correctly, NPV and I must refer to the same functional unit and be calculated over the same period. By using different LCA indicators, it is possible to create specific eco-efficiencies for each indicator. This can be useful to single out and communicate a specific environmental impact in the context of an LCA.

Making use of eco-efficiency opens up many opportunities for comparing technologies, product systems, and services by allowing for a quantitative connection between economic and ecological impacts. In practice, eco-efficiency calculations are used following the ISO 14045 standard by chemical companies such as BASF (BASF, 2018).

While uncertainties in the economic outlook of a product or process are included in the NPV calculation through the discount rate, uncertainties in the LCA's calculation of impacts (as well as consideration of risks with a low likelihood of occurrence) are not always extensively considered in available LCA impact methods. Values of eco-efficiency for specific products or processes should not be considered absolute, and instead rather used only for supporting design comparisons and understanding sensitivities. The development of the eco-efficiency calculation has been important in helping to better take into account the external costs of a product or process. However, a significant and remaining challenge is the fair translation of calculated LCA impacts into monetary values that could be used to reduce the NPV. If and how to place a monetary value on the health of humans and ecosystems is to some extent subjective and, more generally, a topic of moral debate.

5.9 Life Cycle Costing

As mentioned at the start of the chapter, a full life cycle sustainability assessment (LCSA) considers not just LCA but also *life cycle costing (LCC)* and social life cycle assessment (SLCA). LCC predates the development of LCA and is an important aspect to consider in parallel to an LCA. LCC is an analysis of the financial flows of a product or process over its life cycle, and it helps to ensure the economic balance by considering all of the costs involved across each step of the entire life cycle. This includes costs that may be attributed to different stakeholders involved at different stages. While LCC is often done by one particular stakeholder to help inform their own decision-making regarding the product or process in question, it is most beneficial when the results can be presented in a transparent way that helps to inform all stakeholders involved.

Calculations within LCC help to compare product and process variants during the design process in terms of their cost-effectiveness. Example costs that might be involved in an LCC analysis of a *product* include:

- Research and development costs
- Production costs
- Distribution costs
- End-of-life costs (e.g., associated with reuse, recycling, and disposal processes)

And costs to consider within a *process* might involve:

- Research and development costs
- Fixed costs such as construction and standard operation of infrastructure and equipment
- Variable costs such as purchasing chemical feedstocks and energy needed for production

A detailed set of guidelines and proper approach to completing an LCC can be found in Swarr et al. (2011a) and Swarr et al. (2011b), and a more general introduction can be found in Jolliet et al. (2015) and Hauschild et al. (2018).

5.10   Social Life Cycle Assessment

The third and final pillar of a full life cycle sustainability assessment (LCSA) is the completion of a *social life cycle assessment (SLCA)*. SLCAs aim to assess the social and socioeconomic impacts of a chemical product or process across its entire life cycle.

While SLCA follows the general guidelines set out in ISO 14040 for the principles and framework of LCA, some aspects of it differ, and no separate set of standards exists. However, guidelines for SLCA have been developed (UNEP, 2009; Benoît et al., 2010), and discussion regarding its effective application is ongoing (Grubert, 2018; Kühnen and Hahn, 2017; Traverso et al., 2018; Di Cesare et al., 2018). Impact categories that could be considered in an SLCA include human rights, working conditions, cultural heritage, and governance. For a deeper introduction to SLCA, consult the information provided by the United Nations Environment Programme's Life Cycle Initiative (UNEP, 2009).

5.11   Looking Ahead in LCA

As LCA continues to develop and become more frequently applied, it will continue to evolve. For example, while the typical use of LCA previously focused on completing assessments of specific products (in particular to support early design decisions), its role has more recently expanded to also provide companies (and even entire countries) with insights into their overall product portfolios, to help them report on key environmental aspects, and to help them assess options to better manage these (Hellweg and Milà i Canals, 2014). LCA has also played a role in the development of several eco-labels and product declarations that help to communicate information to consumers about the environmental impact (European Commission, 2020; Blue Angel, 2020; Nordic Ecolabel, 2020; EcoLeaf, 2019).

Efforts have also increased to help make regional LCAs possible, which provide insights into geographic differences and allow for comparison of the impacts caused by an activity depending on its location. However, there is still a lack of adequate data describing the emissions and resource flows in different areas, and questions exist about which level of spatial and temporal resolution is sufficient.

In addition to the progress made, there are also a few ongoing challenges still to be addressed. For example, there remains to be a lack of harmonized life cycle data for inclusion in a product's labeling, as well as a shortfall of simple, standardized ways in which life cycle information can be shared with consumers so they understand. This is, of course, in addition to all the data gaps that still exist to describe the emissions and impacts involved across the life cycles of many products.

Room for improvement also still exists for inventory databases, which do not yet share standardized guidelines for ensuring data quality. Many of these inventory databases are also still operating on a subscription basis. Ideally, life cycle inventory information would instead be available for open and public use; however, this is being hindered in part by the significant costs still associated with developing and maintaining such databases.

Questions also continue to remain about how much uncertainty is acceptable within an LCA. This is challenging given that life cycle data for parts of the supply chain are often still unknown, even to the companies creating the products. While manufacturers may sometimes have contact with their direct suppliers, information from further up in the value chain may have already been lost or not even tracked to begin with. New technologies and concepts such as blockchain are now being discussed as possible solutions to help close such data gaps and better manage data (Zhang et al., 2020; Bumblauskas et al., 2019; Bjerkenes and Haddara, 2020).

References

Althaus HJ, Chudacoff M, Hischier R, Jungbluth N, Osses M, Primas A (2007) Life Cycle Inventories of Chemicals. Tech. rep., Swiss Centre for Life Cycle Inventories, Dübendorf

Bare J (2011) TRACI 2.0: the tool for the reduction and assessment of chemical and other environmental impacts 2.0. Clean Technologies and Environmental Policy 13(5):687–696, https://doi.org/10.1007/s10098-010-0338-9, URL http://link.springer.com/10.1007/s10098-010-0338-9

BASF (2018) Eco-Efficiency Analysis. URL https://www.basf.com/en/company/sustainability/management-and-instruments/quantifying-sustainability/eco-efficiency-analysis.html

Benoît C, Norris GA, Valdivia S, Ciroth A, Moberg A, Bos U, Prakash S, Ugaya C, Beck T (2010) The guidelines for social life cycle assessment of products: just in time! The International Journal of Life Cycle Assessment 15(2):156–163, https://doi.org/10.1007/s11367-009-0147-8, URL http://link.springer.com/10.1007/s11367-009-0147-8

Bjerkenes M, Haddara M (2020) Blockchain Technology Solutions for Supply Chains. Springer International Publishing, pp 909–918, https://doi.org/10.1007/978-3-030-32520-6_65, URL http://link.springer.com/10.1007/978-3-030-32520-6_65

Blue Angel (2020) Survey of all Basic Award Criteria. URL https://www.blauer-engel.de/en/companies/basic-award-criteria

Brealey R, Myers S, Allen F (2017) Principles of Corporate Finance, 12th edn. McGraw-Hill

Bulle C, Margni M, Patouillard L, Boulay AM, Bourgault G, De Bruille V, Cao V, Hauschild M, Henderson A, Humbert S, Kashef-Haghighi S, Kounina A, Laurent A, Levasseur A, Liard G, Rosenbaum RK, Roy PO, Shaked S, Fantke P, Jolliet O (2019) IMPACT World+: a globally regionalized life cycle impact assessment method. The International Journal of Life Cycle Assessment 24(9):1653–1674, https://doi.org/10.1007/s11367-019-01583-0, URL http://link.springer.com/10.1007/s11367-019-01583-0

Bumblauskas D, Mann A, Dugan B, Rittmer J (2019) A blockchain use case in food distribution: Do you know where your food has been? International Journal of Information Management (March):102,008, https://doi.org/10.1016/j.ijinfomgt.2019.09.004, URL https://linkinghub.elsevier.com/retrieve/pii/S026840121930461X

Capello C, Fischer U, Hungerbühler K (2007) What is a green solvent? A comprehensive framework for the environmental assessment of solvents. Green Chemistry 9(9):927, https://doi.org/10.1039/b617536h, URL http://xlink.rsc.org/?DOI=b617536h

De Schryver AM, Humbert S, Huijbregts MAJ (2013) The influence of value choices in life cycle impact assessment of stressors causing human health damage. The International Journal of Life Cycle Assessment 18(3):698–706, https://doi.org/10.1007/s11367-012-0504-x, URL http://link.springer.com/10.1007/s11367-012-0504-x

Derwent R, Jenkin M, Passant N, Pilling M (2007) Photochemical ozone creation potentials (POCPs) for different emission sources of organic compounds under European conditions estimated with a Master Chemical Mechanism. Atmospheric Environment 41(12):2570–2579, https://doi.org/10.1016/j.atmosenv.2006.11.019, URL https://linkinghub.elsevier.com/retrieve/pii/S135223100601140X

Di Cesare S, Silveri F, Sala S, Petti L (2018) Positive impacts in social life cycle assessment: state of the art and the way forward. The International Journal of Life Cycle Assessment 23(3):406–421, https://doi.org/10.1007/s11367-016-1169-7, URL http://link.springer.com/10.1007/s11367-016-1169-7

EcoLeaf (2019) EcoLeaf Environmental Label. URL http://www.ecoleaf-jemai.jp/eng/

Ellen Macarthur Foundation (2019) Concept. URL https://www.ellenmacarthurfoundation.org/circular-economy/concept

European Commission (2020) EU Ecolabel. URL https://ec.europa.eu/environment/ecolabel/

Frischknecht R, Büsser Knöpfel S (2013) Swiss Eco-Factors 2013 according to the Ecological Scarcity Method: Methodological fundamentals and their application in Switzerland. Tech. rep., Swiss Federal Office for the Environment, URL https://www.bafu.admin.ch/bafu/en/home/topics/economy-consumption/economy-and-consumption--publications/publications-economy-and-consumption/eco-factors-2015-scarcity.html

Frischknecht R, Wyss F, Büsser Knöpfel S, Lützkendorf T, Balouktsi M (2015) Cumulative energy demand in LCA: the energy harvested approach. International Journal of Life Cycle Assessment 20(7):957–969, https://doi.org/10.1007/s11367-015-0897-4, URL http://dx.doi.org/10.1007/s11367-016-1073-1

GreenDelta GmbH (2019) openLCA Nexus. URL https://nexus.openlca.org/

Gregory JR, Montalbo TM, Kirchain RE (2013) Analyzing uncertainty in a comparative life cycle assessment of hand drying systems. The International Journal of Life Cycle Assessment 18(8):1605–1617, https://doi.org/10.1007/s11367-013-0606-0, URL http://link.springer.com/10.1007/s11367-013-0606-0

Grubert E (2018) Rigor in social life cycle assessment: improving the scientific grounding of SLCA. The International Journal of Life Cycle Assessment 23(3):481–491, https://doi.org/10.1007/s11367-016-1117-6, URL http://link.springer.com/10.1007/s11367-016-1117-6

Hauschild MZ, Huijbregts MA (eds) (2015) Life Cycle Impact Assessment. LCA Compendium - The Complete World of Life Cycle Assessment, Springer Netherlands, Dordrecht, https://doi.org/10.1007/978-94-017-9744-3, URL http://link.springer.com/10.1007/978-94-017-9744-3

Hauschild MZ, Huijbregts M, Jolliet O, Macleod M, Margni M, van de Meent D, Rosenbaum RK, McKone TE (2008) Building a Model Based on Scientific Consensus for Life Cycle Impact Assessment of Chemicals: The Search for Harmony and Parsimony. Environmental Science & Technology 42(19):7032–7037, https://doi.org/10.1021/es703145t, URL https://pubs.acs.org/doi/10.1021/es703145t

Hauschild MZ, Rosenbaum RK, Olsen SI (eds) (2018) Life Cycle Assessment. Springer International Publishing, Cham, https://doi.org/10.1007/978-3-319-56475-3, URL http://link.springer.com/10.1007/978-3-319-56475-3

Hellweg S, Milà i Canals L (2014) Emerging approaches, challenges and opportunities in life cycle assessment. Science 344(6188):1109–1113, https://doi.org/10.1126/science.1248361, URL https://science.sciencemag.org/content/344/6188/1109

Henderson AD, Hauschild MZ, van de Meent D, Huijbregts MAJ, Larsen HF, Margni M, McKone TE, Payet J, Rosenbaum RK, Jolliet O (2011) USEtox fate and ecotoxicity factors for comparative assessment of toxic emissions in life cycle analysis: sensitivity to key chemical properties. The International Journal of Life Cycle Assessment 16(8):701–709, https://doi.org/10.1007/s11367-011-0294-6, URL http://link.springer.com/10.1007/s11367-011-0294-6

Hermans C, Erickson J (2007) Multicriteria Decision Analysis: Overview and Implications for Environmental Decision Making. In: Ecological Economics of Sustainable Watershed Management, Elsevier, pp 213–228, https://doi.org/10.1016/S1569-3740(07)07010-1, URL https://www.emeraldinsight.com/10.1016/S1569-3740(07)07010-1

Hischier R (2007) Life Cycle Inventories of Packaging and Graphical Paper. Tech. rep., Swiss Centre for Life Cycle Inventories, Dübendorf

Huijbregts MAJ, Rombouts LJA, Hellweg S, Frischknecht R, Hendriks AJ, van de Meent D, Ragas AMJ, Reijnders L, Struijs J (2006) Is Cumulative Fossil Energy Demand a Useful Indicator for the Environmental Performance of Products? Environmental Science & Technology 40(3):641–648, https://doi.org/10.1021/es051689g, URL https://pubs.acs.org/doi/10.1021/es051689g

Huijbregts MAJ, Hellweg S, Frischknecht R, Hendriks HWM, Hungerbühler K, Hendriks AJ (2010) Cumulative Energy Demand As Predictor for the Environmental Burden of Commodity Production. Environmental Science & Technology 44(6):2189–2196, https://doi.org/10.1021/es902870s, URL https://pubs.acs.org/doi/10.1021/es902870s

Huijbregts MAJ, Steinmann ZJN, Elshout PMF, Stam G, Verones F, Vieira M, Zijp M, Hollander A, van Zelm R (2017) ReCiPe2016: a harmonised life cycle impact assessment method at midpoint and endpoint level. The International Journal of Life Cycle Assessment 22(2):138–147, https://doi.org/10.1007/s11367-016-1246-y, URL http://link.springer.com/10.1007/s11367-016-1246-y

ISO (2006a) ISO 14040:2006(EN) Environmental management - Life cycle assessment - Principles and framework. Tech. rep., URL https://www.iso.org/standard/37456.html

ISO (2006b) ISO 14044:2006(EN) Environmental management - Life cycle assessment - Requirements and guidelines. Tech. rep., URL https://www.iso.org/standard/38498.html

ISO (2012) ISO 14045:2012(EN) Environmental management - Eco-efficiency assessment of product systems - Principles, requirements and guidelines. Tech. rep., URL https://www.iso.org/standard/43262.html

Jolliet O, Saade-Sbeih M, Shaked S, Jolliet A, Crettaz P (2015) Environmental Life Cycle Assessment. CRC Press, URL https://www.crcpress.com/Environmental-Life-Cycle-Assessment/Jolliet-Saade-Sbeih-Shaked-Jolliet-Crettaz/p/book/9781439887660

Kühnen M, Hahn R (2017) Indicators in Social Life Cycle Assessment: A Review of Frameworks, Theories, and Empirical Experience. Journal of Industrial Ecology 21(6):1547–1565, https://doi.org/10.1111/jiec.12663, URL http://doi.wiley.com/10.1111/jiec.12663

Laurent A, Lautier A, Rosenbaum RK, Olsen SI, Hauschild MZ (2011) Normalization references for Europe and North America for application with USEtox characterization factors. The International Journal of Life Cycle Assessment 16(8):728–738, https://doi.org/10.1007/s11367-011-0285-7, URL http://link.springer.com/10.1007/s11367-011-0285-7

Lloyd SM, Ries R (2008) Characterizing, Propagating, and Analyzing Uncertainty in Life-Cycle Assessment: A Survey of Quantitative Approaches. Journal of Industrial Ecology 11(1):161–179, https://doi.org/10.1162/jiec.2007.1136, URL http://doi.wiley.com/10.1162/jiec.2007.1136

Myhre G, Shindell D, Bréon FM, Collins W, Fuglestvedt J, Huang J, Koch D, Lamarque JF, Lee D, Mendoza B, Nakajima T, Robock A, Stephens G, Takemura T, Zhan H (2013) Anthropogenic and Natural Radiative Forcing. In: Stocker T, Qin D, Plattner GK, Tignor M, Allen S, Boschung J, Nauels A, Xia Y, Bex V, Midgle P (eds) Climate Change 2013: The Physical Science Basis. Contribution of Working Group I to the Fifth Assessment Report of the Intergovernmental Panel on Climate Change, Cambridge University Press, Cambridge, UK and New York, USA, pp 659–740, URL https://www.ipcc.ch/site/assets/uploads/2018/02/WG1AR5_Chapter08_FINAL.pdf

Nordic Ecolabel (2020) Life cycle perspective. URL https://www.nordic-ecolabel.org/why-choose-ecolabelling/life-cycle/

NREL (2012) U.S. Life Cycle Inventory Database. URL https://www.nrel.gov/lci/

Pizzol M, Laurent A, Sala S, Weidema B, Verones F, Koffler C (2017) Normalisation and weighting in life cycle assessment: quo vadis? The International Journal of Life Cycle Assessment 22(6):853–866, https://doi.org/10.1007/s11367-016-1199-1, URL http://link.springer.com/10.1007/s11367-016-1199-1

RIVM (2017) ReCiPe 2016 v1.1. Tech. rep., URL www.rivm.nl/en

Rosenbaum RK, Bachmann TM, Gold LS, Huijbregts MAJ, Jolliet O, Juraske R, Koehler A, Larsen HF, MacLeod M, Margni M, McKone TE, Payet J, Schuhmacher M, van de Meent D, Hauschild MZ (2008) USEtox–the UNEP-SETAC toxicity model: recommended characterisation factors for human toxicity and freshwater ecotoxicity in life cycle impact assessment. The International Journal of Life Cycle Assessment 13(7):532–546, https://doi.org/10.1007/s11367-008-0038-4, URL http://link.springer.com/10.1007/s11367-008-0038-4

Sleeswijk AW, van Oers LF, Guinée JB, Struijs J, Huijbregts MA (2008) Normalisation in product life cycle assessment: An LCA of the global and European economic systems in the year 2000. Science of The Total Environment 390(1):227–240, https://doi.org/10.1016/j.scitotenv.2007.09.040, URL https://linkinghub.elsevier.com/retrieve/pii/S0048969707010522

Swarr TE, Hunkeler D, Klöpffer W, Pesonen HL, Ciroth A, Brent AC, Pagan R (2011a) Environmental life-cycle costing: a code of practice. The International Journal of Life Cycle Assessment 16(5):389–391, https://doi.org/10.1007/s11367-011-0287-5, URL http://link.springer.com/10.1007/s11367-011-0287-5

Swarr TE, Hunkeler D, Klopffer W, Pesonen HL, Ciroth A, Brent AC, Pagan R (eds) (2011b) Environmental Life Cycle Costing: A Code of Practice. Society of Environmental Toxicology and Chemistry, URL https://www.setac.org/store/ViewProduct.aspx?id=1033860

Traverso M, Bell L, Saling P, Fontes J (2018) Towards social life cycle assessment: a quantitative product social impact assessment. The International Journal of Life Cycle Assessment 23(3):597–606, https://doi.org/10.1007/s11367-016-1168-8, URL http://link.springer.com/10.1007/s11367-016-1168-8

Trenberth KE, Fasullo JT, Kiehl J (2009) Earth's Global Energy Budget. Bulletin of the American Meteorological Society 90(3):311–324, https://doi.org/10.1175/2008BAMS2634.1, URL http://journals.ametsoc.org/doi/10.1175/2008BAMS2634.1

UNEP (2009) Guidelines for Social Life Cycle Assessment of Products. URL https://www.lifecycleinitiative.org/starting-life-cycle-thinking/life-cycle-approaches/social-lca/

UNEP (2016) Global Guidance for Life Cycle Impact Assessment Indicators Volume 1. Tech. rep., URL https://www.lifecycleinitiative.org/training-resources/global-guidance-lcia-indicators-v-1/

UNEP (2019a) Global Guidance for Life Cycle Impact Assessment Indicators Volume 2. Tech. rep., URL https://www.lifecycleinitiative.org/training-resources/global-guidance-for-life-cycle-impact-assessment-indicators-volume-2/

UNEP (2019b) Interactive map of LCA databases. URL https://www.lifecycleinitiative.org/applying-lca/lca-databases-map/

UNEP (2020) Life Cycle Initiative. URL https://www.lifecycleinitiative.org/

Verones F, Bare J, Bulle C, Frischknecht R, Hauschild M, Hellweg S, Henderson A, Jolliet O, Laurent A, Liao X, Lindner JP, Maia de Souza D, Michelsen O, Patouillard L, Pfister S, Posthuma L, Prado V, Ridoutt B, Rosenbaum RK, Sala S, Ugaya C, Vieira M, Fantke P (2017) LCIA framework and cross-cutting issues guidance within the UNEP-SETAC Life Cycle Initiative. Journal of Cleaner Production 161:957–967, https://doi.org/10.1016/j.jclepro.2017.05.206, URL https://linkinghub.elsevier.com/retrieve/pii/S0959652617311587

Verones F, Huijbregts MA, Azevedo LB, Chaudhary A, Cosme N, de Baan L, Fantke P, Hauschild M, Henderson AD, Jolliet O, Mutel CL, Owsianiak M, Pfister S, Preiss P, Roy PO, Scherer L, Steinmann Z, Van Zelm R, Van Dingenen R, van Goethem T, Vieira M, Hellweg S (2019) LC-IMPACT Version 1.0: A spatially differentiated life cycle impact assessment approach. Tech. rep., https://doi.org/10.5281/zenodo.3663305, URL https://zenodo.org/record/3663305

Weidema BP, Wesnæs MS (1996) Data quality management for life cycle inventories - an example of using data quality indicators. Journal of Cleaner Production 4(3-4):167–174, https://doi.org/10.1016/S0959-6526(96)00043-1, URL https://doi.org/10.1016/S0959-6526(96)00043-1

Wernet G, Bauer C, Steubing B, Reinhard J, Moreno-Ruiz E, Weidema B (2016) The ecoinvent database version 3 (part I): overview and methodology. The International Journal of Life Cycle Assessment 21(9):1218–1230, https://doi.org/10.1007/s11367-016-1087-8, URL https://doi.org/10.1007/s11367-016-1087-8

Zanghelini GM, Cherubini E, Soares SR (2018) How Multi-Criteria Decision Analysis (MCDA) is aiding Life Cycle Assessment (LCA) in results interpretation. Journal of Cleaner Production 172:609–622, https://doi.org/10.1016/j.jclepro.2017.10.230, URL https://linkinghub.elsevier.com/retrieve/pii/S0959652617325350

Zhang A, Zhong RY, Farooque M, Kang K, Venkatesh VG (2020) Blockchain-based life cycle assessment: An implementation framework and system architecture. Resources, Conservation and Recycling 152, https://doi.org/10.1016/j.resconrec.2019.104512, URL https://linkinghub.elsevier.com/retrieve/pii/S0921344919304185

Risk Assessment and Management of Chemical Products   6

6.1   The Diversity of Chemical Products and Their Risks

With such a wide range of structural variations available on the molecular level, there is of course a very diverse set of chemicals available on the global market today. An ever-increasing number of them—just over 22,000 and 45,000 industrial chemicals are registered and commercially available in the EU and the United States, respectively (ECHA, 2020; US EPA, 2019)—are widely used in industry, commercial, and consumer products, for example, as:

- Biologically active agents such as pharmaceuticals or pesticides
- Polymeric materials such as epoxy resins, polyesters, etc. used, e.g., in packaging and construction
- Industrial chemicals such as dyes, additives, detergents, solvents, etc.

A review of existing inventories internationally even estimates the number of registered chemicals globally to be as high as 350,000 (Wang et al., 2020). Manufactured chemicals can be marketed worldwide, and their annual production volumes can vary widely, from less than 100 kg per year up to 100,000 tonnes per year and more. They can also be combined together in *mixtures* or *formulations* that have targeted properties or serve specific functions different from those a single chemical may have on its own. However, in this book, the focus is placed on understanding *chemical products* as the creation and use of single chemicals.

While chemical products are designed to meet a specific need or perform a unique function, the quality of a chemical product is also evaluated by considering the risks it poses to humans and the environment as undesired side effects. Understanding and minimizing (or preferably avoiding) these risks are where product risk assessment plays an important role, and it is complementary to the use of life cycle assessment (see Chap. 5). *Synthetic chemicals* are those that are manufactured by humans; they

K. Hungerbühler et al., *Chemical Products and Processes*,
https://doi.org/10.1007/978-3-030-62422-4_6

are often different from chemicals that occur naturally in the environment and therefore likely to have increased potential to pose a risk to natural systems.

As a method, *product risk assessment* aims to avoid damage to humans and the environment by characterizing and managing the hazardous properties of and exposure to a chemical product. This chapter on product risk assessment focuses on the *toxic* effects of a product. In the next chapter on *process risk assessment* (Chap. 7), risks posed by *physicochemical* effects are investigated, which are connected with hazardous properties relating to the release of chemical energy (such as fire and explosion).

In view of the multiple hazards that can arise from a new chemical product, *prospective* product risk assessment aims to prevent, both predictively and systematically, the use of a chemical product that causes harm to humans and/or the environment. The goal here is to integrate the safety of a product into its development, also in early design stages.

Product safety comprises (1) knowledge of the relevant product properties that pose a hazard to humans or the environment and (2) knowledge of the measures necessary for safe application, use, and disposal. Furthermore, it deals with two main groups of products: *technical products* that are used within industry and handled by industry employees and *consumer products* that are used within the household and handled by consumers. Chemicals within both groups of products can also enter the environment, meaning that both human health risks and environmental risks need to be assessed.

In contrast to the more general definition of risk that considers the probability of occurrence and extent of damage, the concept of risk assessment involved here does not always explicitly consider probability. Instead, both the description of exposures and the resulting effects are often deterministic. However, existing uncertainties can still be considered by including confidence factors, and some probabilistic risk assessment methods exist and will be introduced.

In addition to the many benefits that it provides when applied voluntarily, the need for product risk assessment also originates from clear regulatory requirements. In the EU (and in many other regions), a new chemical product must be declared to and registered with the appropriate authority before it can be marketed (see Chap. 3). The time required for this registration often determines the time to market—a time that today more than ever is a key factor for the competitiveness and market success of a new product.

In this chapter, the concept of product risk assessment as well as its associated opportunities and limitations is presented and discussed. The information provided here refers to a framework that is largely based on EU legislation and focuses on the development of new chemical products. A case study example using cotton dyeing is provided to illustrate the steps within a product risk assessment. Further, detailed information and illustrative examples of applying risk assessment of chemical products can be found in van Leeuwen and Vermeire (2007) and ECHA (2018).

6.2   Procedure of Product Risk Assessment

Similar to LCA, product risk assessment follows a set procedure with a few key steps as outlined in the flowchart shown in Fig. 6.1. It can, however, also vary slightly in the level of detail involved depending on whether the product is a new or existing substance and whether an initial assessment limited to a set of specific protection goals is needed or a more comprehensive assessment is required. Results from the assessment can be the basis for reducing risks in the next iteration of the R&D cycle and in production, insurance (product liability), marketing, or public relations. An officially recognized risk assessment can also serve as the basis for complying with legal regulations on the distribution and use of the product.

The principle of product risk assessment is classically based on a comparison of the concentrations to which the protected good (humans or the environment) is exposed and the concentration at which an adverse effect occurs.

As the first step of a product risk assessment, *hazard identification* aims to determine the adverse effects that a substance is inherently able to cause. Subsequently, *exposure assessment* focuses on identifying the most critical emission and dispersion scenarios during the use of a product and using them to calculate the concentrations that humans and the environment are exposed to. An *effect assessment* is also carried out to investigate the harmful effects the product can cause, with the aim to determine the relationship between the dose and severity of the adverse effects. The next step of *risk characterization* combines the results of the exposure and effect assessments to characterize the specific risk the product poses by comparing the predicted level of exposure with the threshold of a harmful effect.

The most common approach to describing this risk is the formation of a *risk quotient* (RQ). For the example of environmental toxicity to an aquatic ecosystem, the risk quotient can be simply defined as:

$$RQ_{aq.ecosystem} = \frac{\text{Level of exposure for aquatic organisms}}{\text{No-Effect level for aquatic organisms}} = \frac{PEC_{aq.}}{PNEC_{aq.}} \quad (6.1)$$

Here, *PEC* stands for "predicted environmental concentration" (from exposure assessment) and *PNEC* means "predicted no-effect concentration" (from effect assessment).

The risk quotient, however, is not an absolute measure of risk, but rather indicates that the possibility of occurrence of adverse effects increases with an increasing *PEC/PNEC* ratio. In this way, risks can be compared, but the actual risk cannot be predicted. Despite this limitation, the risk quotient is by far the most widely recognized risk assessment method internationally.

With the risk characterized, it can be evaluated in terms of data quality, protection goals, risk-benefit ratio, societal acceptance, etc. in order to classify it as being either acceptable or not acceptable. If not acceptable, the *risk management* step then identifies risk reduction measures that need to be taken to ensure an acceptable level of residual risk. *Monitoring* serves as an additional component within this step

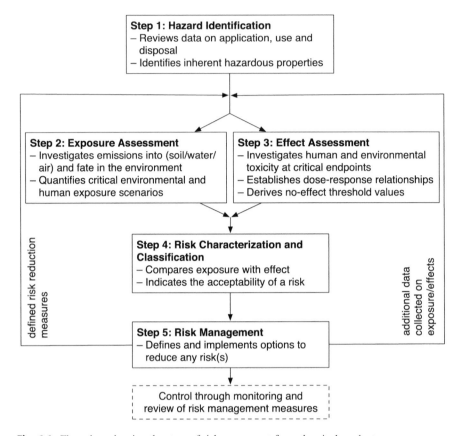

Fig. 6.1 Flow chart showing the steps of risk assessment for a chemical product

and aims to check the effectiveness of the safety measures put in place and detect any unacceptable changes in human or environmental health at an early stage.

The following sections introduce each of the steps of product risk assessment and provide an overview of their fundamental concepts and relevant equations.

6.3   Hazard Identification (Step 1)

The first step of the risk assessment involves identifying adverse effects that a product has an inherent ability to cause. This step directly addresses the second principle of integrated development: *inherent safety* (see Sect. 4.3.2). This is done by collecting information on the types of effects that could occur to humans (such as disease) or to the environment (such as harm to biota) as a result of exposure to the product. Reviewing previous studies and accident records or running preliminary toxicity tests can help identify these hazards. An overview of common hazards to consider is provided here for human toxicity (Sect. 6.3.1) and for environmental toxicity (Sect. 6.3.2).

6.3.1 Human Toxicity

A risk assessment of a product for adverse effects on humans aims to clarify any potential toxic effects during the life cycle of the product for directly exposed populations, such as workers and consumers. In addition to direct exposure, possible effects due to indirect environmental exposure (i.e., through food, drinking water, and respiratory air) are also a target of the investigation, with the inhalation and oral and dermal exposure routes considered.

Identified hazards to human health often include the following toxicity types (van Leeuwen and Vermeire, 2007; ECHA, 2011):

- *Acute toxicity:* A substance's ability to cause adverse effects on human health that occur very soon after exposure (on the order of seconds, minutes, hours, or days).
- *Sub-chronic or chronic toxicity:* A substance's ability to cause adverse effects that occur following long-term, repeated exposure. Sub-chronic effects occur from repeated exposure over several weeks or months, and chronic effects occur from exposure over many months or years. (Note: In general, the definition of "chronic" is in relationship to the lifetime of the organisms of the species considered.)
- *Mutagenicity:* Mutagenic substances can cause permanent, transmissible changes (mutations) in the genetic material of a cell.
- *Carcinogenicity:* Carcinogenic substances can act as an initiator or promoter of abnormal cell growth and differentiation that can lead to cancer. Two mechanisms that have been distinguished for assessment of this are (1) genotoxic carcinogens that cause cancer as a result of direct contact to DNA and (2) non-genotoxic carcinogens that cause their carcinogenic effects not via the DNA, but, for example, via specific receptors.
- *Reproductive toxicity:* This term summarizes adverse effects on a developing embryo or fetus caused either by a chemical directly acting on its cells or by a chemical inducing mutation in a parent's germ cell.
- *Irritation and corrosivity:* Irritating substances can cause inflammation when in contact with certain tissues, e.g., skin, eyes, or mucous membranes. Corrosive substances can destroy living tissue they contact.
- *Sensitization:* Sensitizers are substances that, when inhaled or absorbed through the skin, cause an allergic reaction and lead to a more pronounced reaction when there is contact with the substance in the future.

A chemical product can be tested for each of these toxicity types. For decades, this has been done often by using standardized animal studies and then using those results to estimate how a human would respond to the same exposure. For acute and chronic toxicity, a wide range of adverse effects could be examined including effects on body weight and various internal organs. Acute toxicity of a substance is often described using the lethal dose (or concentration) of the product that kills 50% of the studied population (abbreviated as LD_{50} or LC_{50}). The concentration that elicits a certain effect on $x\%$ of the studied population can also be determined and communicated as the effect concentration (EC_x). For chronic toxicity, a first estimate is commonly made using standardized 28-day or 90-day tests with rats. These testing methods and effects are further reviewed within the *effect assessment* step (see Sect. 6.5).

Following the Globally Harmonized System (GHS) of Classification and Labelling of Chemicals (UN, 2019), which was introduced in Sect. 3.3.2, Fig. 6.2 illustrates how a substance's acute toxicity is categorized into different levels of hazard for clear and consistent communication to a product's user. The GHS uses a series of signal words, hazard statements, and visual symbols that can be displayed on the product packaging to clearly and consistently communicate hazards to users.

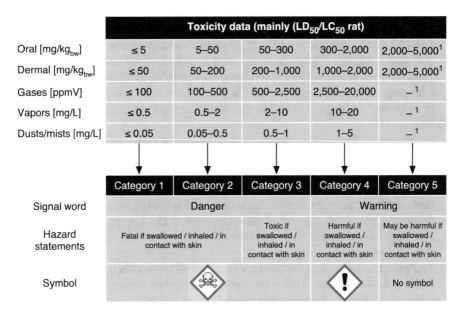

Fig. 6.2 Acute toxicity data resulting in the identification of hazard categories for human health following the Globally Harmonized System (GHS) (UN, 2019); bw = body weight. 1: LD_{50} or LC_{50} values for substances in Category 5 are largely estimated or extrapolated since animal testing in these ranges is discouraged to protect animal welfare. For more information on Category-5 substances, see section 3.1.2 in UN (2019)

6.3.2   Environmental Toxicity

While risk assessment in regard to human health aims to protect everyone (including the most vulnerable), a risk assessment to protect the environment rather aims to protect entire ecosystems and simply cannot test each of the millions of species of fauna and flora that exist on Earth. Instead, a primary estimation of the environmental toxicity of a chemical is often carried out by testing aquatic species. In particular, testing is frequently mandated on algae, daphnia, and fish, which represent three distinct trophic levels.

Adverse effects to screen for in the environment can include, for example, reduction of an organism's survival, growth, or reproduction rates. Environmental toxicity can be categorized into:

- *Aquatic toxicity:* Includes toxic effects on aquatic ecosystems in, for example, rivers, lakes, and oceans. Normally used as the first proxy measure for general environmental toxicity.
- *Sediment toxicity:* Includes toxic effects on organisms that exist within the sediment.
- *Terrestrial toxicity:* Covers toxic effects on terrestrial animals, plants, and microorganisms.
- *Technical ecosystem toxicity:* Includes toxic effects on man-made ecosystems, primarily concerning the health of microorganisms in biological wastewater treatment plants. Toxic effects on such microorganisms can affect the cleaning performance of the plant and therefore pose a risk to the natural environment.

Just as for human toxicity, environmental toxicity includes both acute and chronic effects. Following the GHS, Fig. 6.3 illustrates how the acute and chronic toxicities of a substance are categorized into different levels of hazard for the aquatic environment depending on data availability and the substance's degradability. In addition to considering available data to determine the LC_{50} or EC_{50} values, the categorization also considers chronic effects described by the lowest concentration at which no effect is observed in the studied species. This is known as the *no observed effect concentration* (*NOEC*). The published guidelines for the GHS provide many more helpful details and specific step-by-step instructions for reaching these categorizations (UN, 2019).

In practice, this testing can be resource-intensive for both human and environmental toxicities, and investment needs to be made to ensure testing creates high-quality data and is thoroughly documented. Inspection has shown that the majority of submitted REACH registration dossiers lack important safety information about the chemicals they describe (ECHA, 2017a, 2019).

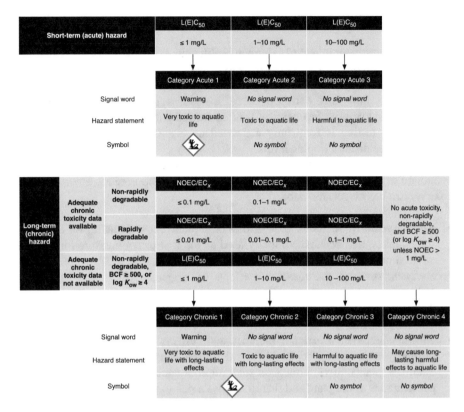

Fig. 6.3 Acute and chronic toxicity data resulting in the identification of hazards to the aquatic environment following the Globally Harmonized System (GHS) (UN, 2019). The criteria for being rapidly degradable are defined in detail in Sect. 4.1.2.11 of the GHS (UN, 2019) and can include, for example, achieving 70% biodegradation within 10 days of the start of degradation. K_{ow}: octanol-water partition coefficient. BCF: bioconcentration factor

6.4 Exposure Assessment (Step 2)

With the hazards of a product identified, this next step aims to identify the level of exposure to the product for humans and the environment. *Exposure* in this context refers to the presence of a chemical that triggers—depending on the type, intensity, and duration of the exposure—a reaction that may be the cause of an adverse effect. The quantitative description of exposure is usually characterized by (1) the *concentration* of a chemical in environmental media, food, or consumer goods, (2) the absorbed *dose* of this chemical by specific organisms, and (3) the *duration* of the exposure. Humans and the environment may be exposed to varying levels of a hazardous chemical at different points along its life cycle. To help simplify the assessment and make it more manageable, exposure assessment prioritizes the investigation of *critical exposure scenarios* that result in the most significant exposures.

Because a risk assessment is proactive in the case of a new chemical, the exposure assessment cannot easily rely on direct measurements. Instead, exposures often have to be estimated by models, and these generally require the following data about the product:

- Amount placed on the market
- Emission potentials during application, use, and disposal
- Biologically available[1] exposure level, e.g., due to physical or chemical product properties, such as particle size, water solubility, etc.
- Frequency and duration of exposure
- Specifically for human exposure:
  - Exposure via air, soil, water, diet, as well as pharmaceutical and personal care products
  - Extent of exposed populations

6.4.1 Human Exposure

Figure 6.4 provides an overview of three categories of human exposure and the different modeling approaches used to understand them. Local, short-term exposures of humans are often caused by process accidents and are largely an occupational safety issue. The dispersion models used to assess these are introduced in more details in Chap. 7. Local, long-term exposures of humans often result from the standard use of products and are a consumer health issue. These can often be modeled with one-compartment models as described further below in this section.

Local, regional, and even global environmental exposures to a chemical are modeled by means of multi-compartment environmental models, and these models are presented in the next section on environmental exposure (Sect. 6.4.2). They capture the situation in which humans are indirectly exposed to the chemicals that circulate in the environment.

Human exposure is an important aspect to carefully consider during the development of a product and to also communicate along the supply chain. As shown in Fig. 6.5, the number of people exposed to a chemical product increases over the course of the product life cycle while at the same time the amount of knowledge those exposed persons have regarding the safe handling of the product decreases.

6.4.1.1 Exposure Routes

Depending on a product's use and properties, different human *exposure routes* (or pathways) exist. The three primary routes of human exposure are oral, inhalation, or dermal, and these are often quantified through calculating the *predicted daily intake* (PDI) as shown in Eq. 6.2. Here, i indicates a contact medium (such as food,

---

[1]Bioavailability: fraction of the total exposure that can readily be absorbed by the particular organism considered and therefore readily interact with its biosystem

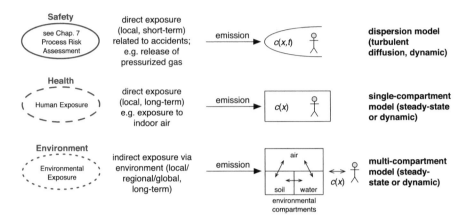

Fig. 6.4 Illustration of safety, health, and environmental exposures of humans and the modeling approaches used for each case

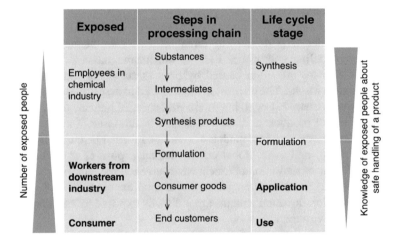

Fig. 6.5 Number of people exposed compared to their knowledge about safe handling of a product over its life cycle

drinking water, or air), and j indicates the exposure route (such as oral, inhalation, or dermal). By taking into account the actual amount of a chemical that can be taken up after an exposure (known as *uptake efficiency*, r_{ij}), the *dose* of the chemical can then be defined as shown in Eq. 6.3. Defining the dose through the oral, inhalation, and dermal exposure routes is illustrated in Fig. 6.6 as defined by the US EPA (1997):

$$PDI_j = \frac{1}{m_{bw}} \sum_i (c_{ij} \times CR_{ij}) \tag{6.2}$$

$$D_j = \frac{1}{m_{bw}} \sum_i (c_{ij} \times CR_{ij} \times r_{ij}) \tag{6.3}$$

- PDI_j: Predicted daily intake for exposure route j [ng kg$_{bw}^{-1}$ d^{-1}].
- D_j: Dose for exposure route j [ng kg$_{bw}^{-1}$ d^{-1}].
- m_{bw}: Human body weight [kg]; a default value of 70 kg is often used.
- c_{ij}: Concentration of chemical in contact medium i through exposure route j. Units are [ng/g food] for oral intake, [ng/m^3 air] for inhalation, and [ng/m^2 skin] for dermal contact.
- $CR_{i,oral}$: Oral intake, for example, according to population-specific diet [g food/d].
- $CR_{i,inhalation}$: Inhalation rate [m^3 air/hr] × exposure time [hr/d] (average inhalation rate: ca. 0.8 m^3/hr).
- $CR_{i,dermal}$: Dermal contact area [m^2 skin] × exposure frequency [d^{-1}] (the maximum contact area of skin is ca. 2 m^2).
- r_{ij}: Chemical-specific uptake efficiency by contact medium i and exposure route j [–]; $0 \leq r \leq 1$.

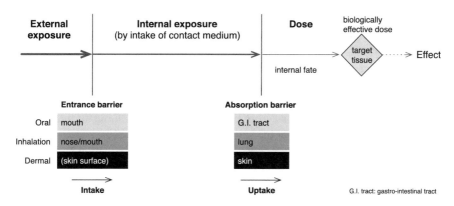

Fig. 6.6 Illustration of the steps from the external exposure to the dose after considering the intake and uptake rates of a chemical into the body

Both the PDI and dose can be summed across all of the different exposure routes (j) to reach a total value.

6.4.1.2 Elimination

Chemicals taken up by the human body are generally metabolized to some extent, often in the liver, before they are excreted with urine and/or feces and, in some cases, also with mother's milk (DDT and other hydrophobic substances, see Gyalpo et al., 2012) and/or menstrual blood (perfluoroalkyl and polyfluoroalkyl substances (PFAS), see Gomis et al., 2017). Generally, the metabolites formed in the body are more hydrophilic (water soluble) than their parent compounds so that they can be excreted more readily. Depending on how rapidly a chemical is modified by the metabolic system of the body, the process of metabolism and excretion takes place on different time scales from rapid to (very) slow.

An example of rapid *elimination* from the human body is *phthalates*, which have been extensively used as plasticizers in plastics such as polyvinyl chloride (PVC). Phthalates are diesters of phthalic acid, such as di-(ethylhexyl) phthalate (DEHP) or dibutyl phthalate (DBP). In the human body, one of the ester bonds is hydrolyzed so that the phthalate monoesters are formed, which are then readily excreted via urine; for the heavier phthalates, also glucuronide conjugates are formed before excretion (Frederiksen et al., 2007). Elimination half-lives of phthalates in humans are on the order of several hours (Koch and Angerer, 2007; Koch et al., 2012).

Much slower metabolism and elimination is found for highly persistent chemicals such as polychlorinated biphenyls (PCBs), in particular the highly chlorinated PCB congeners. The metabolism of PCBs in the human body is complex and only partly understood (Grimm et al., 2015). Ritter et al. (2011) determined elimination half-lives of PCBs from extensive data sets of PCB levels measured in human blood and adipose tissue (without consideration of the mechanisms of PCB metabolism) and found elimination half-lives from 2.6 years (PCB-52) to 15.5 years (PCB-170). Their analysis of PCB concentrations in human tissue shows that there is an important distinction between *apparent* and *intrinsic* elimination half-lives. Apparent half-lives can be determined from the time trends of PCBs in individuals, but if exposure to PCBs of these individuals continues (which is generally the case because of remaining PCB background contamination), the time trend will be (very) shallow and the apparent elimination half-lives very long. Only when the effect of ongoing exposure is removed from the data the intrinsic elimination half-lives, which describe the time scale of the actual elimination process, can be determined. (The half-lives reported by Ritter et al. (2011) are intrinsic elimination half-lives.)

In pharmacokinetic models like the one presented in the next subsection, intrinsic elimination half-lives are needed as input parameters.

6.4.1.3 Example: Exposure via the Diet

To provide an example of human exposure modeling, the dietary (oral) and inhalation exposures to a hydrophobic chemical such as DDT or PCBs are calculated by means of simple, one-compartment pharmacokinetic models.

Through such a model, the oral exposure of a chemical via the diet can be converted into the concentration of the chemical stored in the body's lipids. Calculating this requires knowing the oral dose and the body's elimination rate constant of the chemical.

If the oral dose is *constant*, the resulting concentration of a chemical in the lipid tissue over time ($c_{lip}(t)$) can be calculated as:

$$\frac{dc_{lip}(t)}{dt} = D_{oral} - k_{elim} \times c_{lip}(t) \tag{6.4}$$

$$D_{oral} = \frac{1}{m_{lip}} \times \sum_i c_i \times CR_i \times r_i \tag{6.5}$$

$$c_{lip}(t) = \frac{1 - e^{-k_{elim} \times t}}{k_{elim}} \times D_{oral} \tag{6.6}$$

Eventually, the concentration in the lipids reaches a steady state ("stst"):

$$c_{lip}^{stst} = \frac{D_{oral}}{k_{elim}} \tag{6.7}$$

- $c_{lip}(t)$: time-dependent internal concentration of chemical in human lipid tissue [ng/kg$_{lip}$]
- c_{lip}^{stst}: internal steady-state concentration of chemical in human lipid tissue [ng/kg$_{lip}$]
- k_{elim}: first-order elimination rate constant [d^{-1}] derived from the intrinsic elimination half-life, $t_{1/2}$, as $k_{elim} = \ln 2/t_{1/2}$
- m_{lip}: total mass of lipid tissue in human (adult average value: ca. 10 kg)
- D_{oral}: oral dose [ng kg$_{lip}^{-1}$ d^{-1}]
- Assumptions: $c_{lip}(t = 0) = 0$, $m_{lip} = $ constant, $D_{oral} = $ constant

A plot of the chemical concentration in the lipid tissue over time (as defined in Eq. 6.6) is shown in Fig. 6.7 for the uptake of a chemical using an assumed constant oral dose (D_{oral}) of 1.2 ng/kg$_{lip}$/d and an assumed elimination rate constant (k_{elim}) of 0.384 d^{-1}.

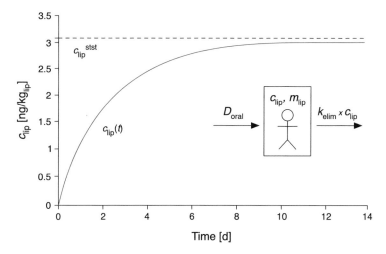

Fig. 6.7 Modeled internal concentration in human lipids (c_{lip}) of a chemical calculated with a one-compartment pharmacokinetic model using a first-order elimination rate constant (k_{elim}) of 0.384 d^{-1} and an oral dose (D_{oral}) of 1.2 ng kg$_{lip}^{-1}$ d^{-1}. A sketch of the simple one-compartment model used is shown within the plot

Once the internal steady-state concentration of the chemical in the lipids is reached, we assume that the oral exposure ends ($D_{oral} = 0$). Then, the body's elimination of the chemical will result in an eventual total removal of the chemical from the lipids. To describe this, Eq. 6.4 can be rewritten as:

$$\frac{dc_{lip}(t)}{dt} = -k_{elim} \times c_{lip}(t) \tag{6.8}$$

where $c_{lip}(t = 0) = c_{lip}^{stst}$. This then results in:

$$c_{lip}(t) = c_{lip}^{stst} \times e^{-k_{elim} \times t} \tag{6.9}$$

Figure 6.8 shows a plot of the resulting chemical concentration in the lipid tissue over time (as defined by Eq. 6.9) using the same elimination rate constant of $0.384\,d^{-1}$ and starting from the steady-state internal concentration of $3.125\,ng/kg_{lip}$.

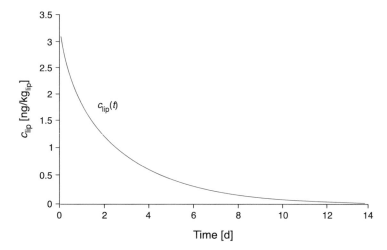

Fig. 6.8 Internal concentration of a chemical in the lipid tissue over time (c_{lip}) modeled with a one-compartment pharmacokinetic model. At $t = 0$, the internal concentration in the body lipids (c_{lip}) is $3.125\,ng/kg_{lip}$, the oral dose (D_{oral}) is zero, and the first-order elimination rate constant (k_{elim}) is $0.384\,d^{-1}$

6.4.1.4 Example: Exposure via Inhalation

A second example is the calculation of human exposure through inhalation of indoor air containing the same hydrophobic chemical. This can be modeled through the use of a simple box model describing a room with an indoor flow of air, a release source of the chemical, and an outflow of air. Combining this with a one-compartment pharmacokinetic model (as done in the previous example) allows yet again for calculation of human exposure expressed through the chemical's concentration in the lipid tissue. Figure 6.9 shows the structure of the box model depicting a single room and a person inhaling air within the room.

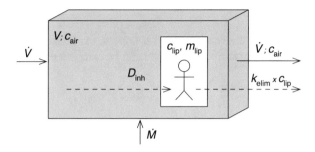

Fig. 6.9 Schematic of a box model used to calculate the concentration of a chemical in indoor air. Within the box model, there is a one-compartment pharmacokinetic model used to calculate the exposure of a human to the chemical through inhalation and expressed as concentration in the body lipids

Variables used within the indoor air box model are defined as follows:

- c_{air}: average indoor concentration [ng/m^3]
- $\dot{M}$: constant release rate of chemical [ng/d]
- V: room volume [m^3]
- $\dot{V}$: ventilation rate [m^3/d]
- τ: indoor residence time of air ($=V/\dot{V}$) [d]

The variables used within the one-compartment pharmacokinetic model again are defined as:

- $c_{lip}(t)$: time-dependent internal concentration of chemical in human lipid tissue [ng/kg$_{lip}$]
- k_{elim}: first-order elimination rate constant [d^{-1}]
- m_{lip}: total mass of lipid tissue in human (adult average value: ca. 10 kg)
- D_{inh}: inhalation dose [ng kg$_{lip}^{-1}$ d^{-1}]
- Q_{inh}: inhalation rate [m^3/d]
- r : uptake efficiency via inhalation [$0 \leq r \leq 1$]
- Assumptions: $c_{lip}(t = 0) = 0$, m_{lip} = constant, D_{inh} = constant

With the model shown in Fig. 6.9, the concentration of the chemical in the air (c_{air}) with a *constant* release rate ($\dot{M}$) can be calculated as:

$$\frac{dc_{air}(t)}{dt} = \frac{\dot{M}}{V} - \frac{\dot{V}}{V} \times c_{air}(t) \qquad (6.10)$$

$$c_{air}(t) = \frac{\dot{M}\tau}{V} \times (1 - e^{-t/\tau}) \qquad (6.11)$$

Eventually, the concentration in air will reach a steady state (c_{air}^{stst}):

$$c_{air}^{stst} = \frac{\dot{M}\tau}{V} \tag{6.12}$$

With the steady-state concentration in air known, the concentration in the body lipids resulting from exposure via inhalation can be calculated as:

$$\frac{dc_{lip}(t)}{dt} = D_{inh} - k_{elim} \times c_{lip}(t) \tag{6.13}$$

$$D_{inh} = \frac{1}{m_{lip}} \times c_{air}^{stst} \times Q_{inh} \times r \tag{6.14}$$

$$c_{lip}(t) = \frac{1 - e^{-k_{elim} \times t}}{k_{elim}} \times D_{inh} \tag{6.15}$$

Eventually, the concentration in the lipids will reach a steady state (c_{lip}^{stst}):

$$c_{lip}^{stst} = \frac{D_{inh}}{k_{elim}} \tag{6.16}$$

Figure 6.10 shows a plot of the resulting chemical concentration in the person's body lipids over time (as defined by Eq. 6.15) with a constant release rate ($\dot{M}$) of 288 ng/d.

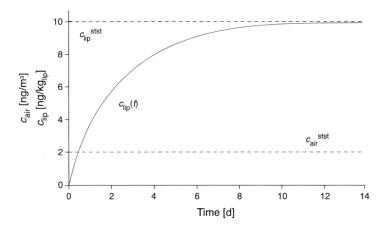

Fig. 6.10 Modeled steady-state indoor air concentration (c_{air}^{stst}), concentration in body lipids (c_{lip}), and steady-state concentration in body lipids (c_{lip}^{stst}) over time derived from a single-box model for indoor air and a one-compartment pharmacokinetic model. This calculation assumes a constant indoor air concentration with a release rate of $\dot{M} = 288$ ng/d, $\tau = 0.278$ d, $V = 40$ m^3, $Q_{inh} = 19.2$ m^3/d, $r = 1$, $m_{lip} = 10$ kg, and $k_{elim} = 0.384$ d^{-1}

While the examples provided here are fairly simple, much more complex models exist for determining the exposures of entire populations considering multiple exposure routes across hundreds of different product types and diets (Trudel et al., 2011; Ritter et al., 2009). For the analysis of indirect exposures via the environment, further models are needed that can describe concentrations and intake via respiratory air, drinking water, food (meat, fish, dairy products, cereals, and vegetables), etc. (Trudel et al., 2011; Ritter et al., 2009).

6.4.2 Environmental Exposure

The exposure assessment step also aims to identify the extent to which the environment is exposed to a chemical. Most commonly, exposures in the environmental compartments of air, water (including sediment), and soil are considered, with the focus often first placed on investigating the water compartment. The starting point for this step is understanding the emissions to each compartment, which can be characterized by the types of sources, e.g., continuous vs. intermittent, the emission rate, or the amounts released, as well as by the chemical's ability to be diluted and potential to react and degrade. Experimentally, the exposure can be quantified through environmental analysis or by using *biomarkers*, which are measurements of biochemical, structural, and functional changes within exposed test organisms. Compartment-specific exposure can be estimated through the calculation of the *predicted environmental concentration* (PEC) by means of models. Units of the PEC are for air mg/m^3, for water mg/L, and for soil mg/kg. The PEC is specific to an environmental compartment. Wherever possible, emissions should be reduced by treatment processes, and treatment processes therefore need to be reflected by the model as well. An overview of emission treatment options and flows throughout the life cycle of a product is shown in Fig. 6.11.

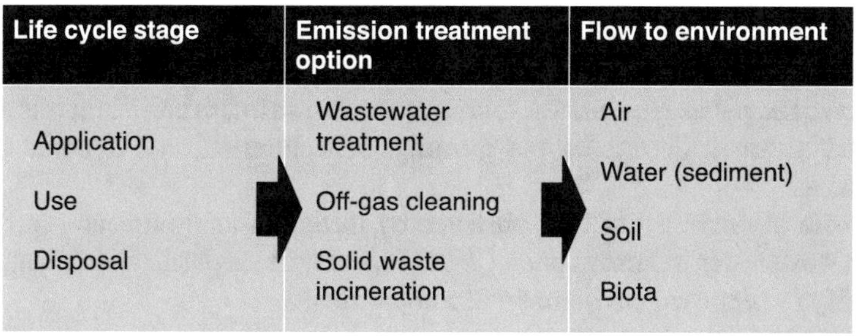

Life cycle stage	Emission treatment option	Flow to environment
Application	Wastewater treatment	Air
Use	Off-gas cleaning	Water (sediment)
		Soil
Disposal	Solid waste incineration	Biota

Fig. 6.11 Overview of common emissions and treatment options for flows into the environment

Depending on the context, the maximum estimated concentrations in the environment (*peak concentrations*) or the average concentrations (*background concentrations*) may be important to calculate.

In the treatment of emissions, care should be taken to avoid shifting a problem from one environmental compartment into another. For example, in a wastewater treatment plant, a problem shift could occur if a substance is not degraded rapidly enough (e.g., where the biodegradation rate constant $k_b < 10^{-3}\,s^{-1}$):

- A shift into exhaust air for volatile substances ($\log K_H > -5$)[2]
- A shift into waste (biomass) for substances with low polarity ($\log K_{ow} > 3$)[3]

Environmental exposure assessment involves using models on two different scales:

- *Local*: Calculation of the local predicted environmental concentrations (PEC_{local}) using emission rates and considering immediate dilution. This includes emissions from point sources (e.g., from a single factory) and from population-related sources (e.g., from municipal wastewater treatment plants).
- *Regional or continental*: Calculation of predicted environmental concentrations for a specific region ($PEC_{regional}$) or an even larger area ($PEC_{continental}$) using emission rates and information about the distribution and transformation of the substance in the environment. This includes emissions from area-related sources (e.g., from large farming operations) or from many point sources that are lumped together.

6.4.2.1 Local Exposure (Point and Diffuse Emission Sources)

Exposure assessment on a local scale focuses on the immediate entry of emissions into the environment and intends to determine the local exposure levels (PEC_{local}).

To estimate local exposures, several models exist with different degrees of accuracy to describe the entry, dilution, and fate of chemicals released into water, air, and soil (van Leeuwen and Vermeire, 2007). Regardless of the exact choice of the model, the following approach can generally be used to estimate the PEC_{local}:

1. Determine the upper and lower bounds of the emission flows and the annual amount of product used.
2. Analyze the potential emission sources and the maximum emission; this analysis should result in describing the quantity, concentration, and dynamics of the emission.
3. Estimate the removal of the substance by means of environmental technology (e.g., wastewater treatment plant (WWTP), exhaust air purification plant, etc.).
4. Identify the exposed environmental compartments.
5. Estimate the amount of substance entering each environmental compartment.

---

[2] K_H: Henry's law constant [Pa m^3 mol^{-1}].
[3] K_{ow}: octanol-water partition coefficient [−]

6. Estimate the environmental concentration at the location of the emission, considering the immediately occurring dilution; these estimates lead to the PEC_{local} in the various compartments.

For existing products, often only sales data are known (e.g., monetary sales figures, customs statistics, etc.), but not data describing the quantities released into the environment. In the case of new products, these data can only be estimated by taking into account the characteristics of application and use.

A simply estimated PEC_{local} is often sufficient in the worst case for an initial assessment of the risk quotient, provided that it is not expected that the substance will be transferred to a different environmental compartment considered more critical to protect. Here, some simple approaches for the calculation of the PEC_{local} for the water compartment are presented through two examples.

The simplest models for estimating the local concentration in water are based on the assumption of the complete mixing of the substance introduced into the body of water. With this approach, the following formulas result for estimating concentrations from point and diffuse sources:

(a) *Point source (e.g., industrial wastewater from a production process):*

$$PEC_{local} = \frac{E\ (1 - f_{ret})}{Q \times D} \qquad (6.17)$$

- PEC_{local}: predicted local environmental concentration in water [kg/m^3]
- E: emission rate [kg/d]
- f_{ret}: fraction retained in WWTP [–]
- Q: effluent flow [m^3/d] (estimate if no data: ca. 10,000 population equivalents = ca. 2000 m^3/d)
- D: dilution factor [–] (value of 10 for release into freshwater environment and of 100 for release into marine environment ECHA, 2016)

(b) *Diffuse source (e.g., consumer products from households):*

$$PEC_{local} = \frac{M \times f_{rel} \times (1 - f_{ret} \times f_{treat})}{P \times W \times D} \qquad (6.18)$$

- PEC_{local}: predicted local environmental concentration in water [kg/m^3]
- M: consumed annual amount of the product [kg/yr]
- f_{rel}: fraction of product released into wastewater (depending on leachability, water solubility, etc.) [–]
- f_{ret}: fraction of product removed in WWTP [–]
- f_{treat}: fraction of wastewater handled by WWTP [–]
- P: population [persons]
- W: wastewater per person and year [m^3 person^{-1} yr^{-1}] (EU: ca. 55–75 m^3 person^{-1} yr^{-1})
- D: dilution factor [–] (value of 10 for release into freshwater environment and of 100 for release into marine environment ECHA, 2016)

Internal exposure of aquatic organisms to a chemical can be quantified by comparing their internal concentrations to the surrounding environmental concentrations as well as to the concentration of the chemical in their food. The *bioconcentration factor* (BCF) is defined as the ratio of the concentration of a chemical in the organism and the chemical's concentration in the surrounding water under the condition that uptake occurs via the gills only, but not via food. This is a setting that can only be realized in laboratory test systems; in the real environment, uptake always occurs via gill respiration and food ingestion in combination. Next, the *biomagnification factor* (BMF) is defined as the ratio of the chemical's concentration in the organism and in the food consumed by the organism. It describes a chemical's ability to be taken up via the food chain in such a way that concentrations in higher trophic levels increase (biomagnification). Finally, the *bioaccumulation factor* (BAF) is defined as the ratio of the concentration of chemical in the organism and the chemical's concentration in the surrounding water, but in this case—in contrast to the BCF—with uptake of the chemical by both gill respiration and food ingestion. The BAF, therefore, considers all exposure routes and multiple physiological processes to describe the net resulting uptake and accumulation of a chemical in an organism. All three factors are further defined in Appendix B.1 and also discussed in the literature (Arnot et al., 2010) and testing standards (OECD, 2012).

6.4.2.2 Regional Exposure (Area-Related Emission Sources)

To estimate exposure on the regional or continental level, the scope of the calculations needs to be greatly expanded. After entering the environment, natural processes of transport, transformation, and accumulation affect the fate of a substance (see Appendices B.1 and B.2 as well as Schwarzenbach et al., 2016). Estimating exposure at this level needs to consider emission characteristics, physicochemical properties of the chemical, and environmental conditions. These considerations can be grouped as questions for analysis into the following dimensions:

- *Compartment:* Does the substance remain in the environmental compartment into which it was emitted? Or does a significant portion move to another compartment?
- *Space:* Does the substance remain in the region where it was emitted? Or is a significant portion transported to other regions?
- *Time:* Is the substance significantly removed in one of the environmental compartments by physical, chemical, or biological processes, making it no longer available for natural cycles and biota? Or is it likely to remain environmentally relevant because of its persistence or its transformation products?
- *Accumulation:* Is the substance significantly enriched in an environmental compartment or in a food chain?

To help answer these questions, *multimedia mass-balance models* are used that describe and estimate the environmental behavior of products and the resulting environmental concentrations in the various compartments. They have the ability to describe the processes of:

- *Mass transport within an environmental compartment:* Through convective mixing, a substance can move according to the flow of the corresponding environmental medium (air, water) within a compartment.
- *Mass transport between environmental compartments:* Through mass flow across media boundaries, the substance passes into other environmental compartments, e.g., by diffusion (evaporation), advective outflow, etc.
- *Substance transformation through biotic degradation:* Substances can be biologically converted, for example, by microorganisms in the soil or water.
- *Substance transformation by abiotic degradation:* Hydrolysis, photolysis, oxidation, etc. can cause chemical transformation.
- *Sorption:* Substances can be sorbed onto particles in soils, waters, and sediments. This leads to them becoming enriched and retained in that compartment.
- *Bioaccumulation:* Substances can accumulate in organisms along the food chain (particularly hydrophobic substances).

As outlined in Fig. 6.12, the modeling of these processes requires three types of data for input into a mass-balance model: emission data, substance data, and environmental data.

The basic equation of the mass balance for a chemical with concentration c in a single, homogeneous environmental compartment with volume V is:

$$V \frac{dc}{dt} = \text{inflow} - \text{outflow} + Q_{\text{emission}} - Q_{\text{transformation}} \tag{6.19}$$

- $V \frac{dc}{dt}$: accumulation [kg/s]
- inflow, outflow: transport processes across media boundaries in the environment (diffusion, advection)
- Q_{emission}: emission inflow (source term in the compartment)
- $Q_{\text{transformation}}$: transformation flow, e.g., degradation

For the simple case that in a single, homogeneously mixed environmental compartment the fate of a chemical is controlled solely by degradation with the degradation rate constant k_{deg} (no input or output), the integration of Eq. 6.19 yields the dynamic concentration curve $c(t)$ defined in Eq. 6.20:

$$\frac{dc(t)}{dt} = \frac{Q_{\text{emission}}}{V} - k_{\text{deg}} \times c(t) \xrightarrow{c(t_0)=0} c(t) = \frac{Q_{\text{emission}}}{k_{\text{deg}} \times V} \left(1 - e^{-k_{\text{deg}} \, t}\right)$$

$$\tag{6.20}$$

When Eq. 6.20 is plotted, Fig. 6.13 shows that a steady-state environmental concentration (c^{stst}) is reached asymptotically and, under these assumptions, is

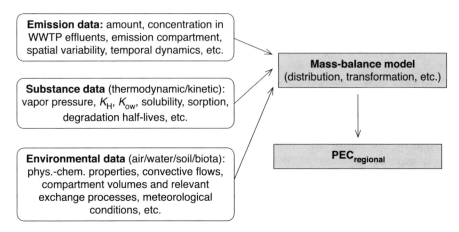

Fig. 6.12 Data inputs needed to estimate the regional or continental predicted environmental concentration (PEC) in the different compartments of a mass-balance environmental fate model

proportional to the emission inflow and inversely proportional to the degradation rate constant:

$$c^{\text{stst}} = \frac{Q_{\text{emission}}}{k_{\text{deg}} \times V} \tag{6.21}$$

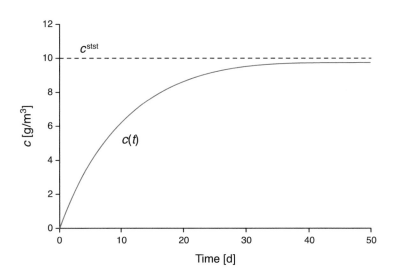

Fig. 6.13 Plot of concentration over time of a substance in a single compartment with only degradation according to Eq. 6.20 to reach a steady-state concentration (c^{stst}) of 10 g/m³. Parameters used: $k_{\text{deg}} = 0.1\,\text{d}^{-1}$, $Q_{\text{emission}} = 100\,\text{g/d}$, $V = 100\,\text{m}^3$

For the general modeling of regional and continental environmental concentrations and the estimation of substance accumulation in selected food chains, the approach of the single-compartment model shown above can be extended to multi-compartment environmental fate models (Di Guardo et al., 2018). Figure 6.14 depicts an example structure of such a model.

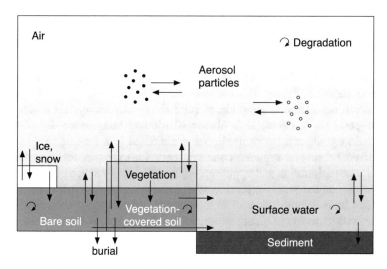

Fig. 6.14 Example structure of a multimedia environmental fate model. Modeled compartments include: air, water, sediments, two soil types, vegetation, and ice/snow

These models are used to provide an estimate of the environmental fate of a chemical and do this by depicting the environment through a set of standardized compartments and with four levels of complexity (Mackay, 2001):

- Level I: A closed system at thermodynamic equilibrium as determined by the partition coefficients between air, water, soil, etc.; no consideration of substance transformation or transport (i.e., there is a single emission of a certain amount of the chemical into the compartment).
- Level II: In addition to level I, consideration of a continuous inflow of emissions and of substance transformation (e.g., degradation). This is an open system at equilibrium.
- Level III: In addition to level II, consideration of transport between environmental compartments. This is an open system at steady state, but not at equilibrium: transport between compartments such as deposition from air to water with falling rain can push the system out of equilibrium.
- Level IV: In addition to level III, consideration of dynamic emissions and calculation of the resulting temporal change of the concentrations in all compartments. This is an open system, not at steady state and not at equilibrium.

Generally, the distribution of a chemical between compartments is described by partition coefficients, and degradation processes are described by degradation rate constants (see Appendix B).

Multi-compartment models as described here can often be run with a relatively small set of chemical property data, e.g., vapor pressure, water solubility, K_{ow}, and $k_{transformation}$. However, it is important to keep in mind that these property data describe the environmental fate of nonionizing organic chemicals. The environmental fate of acids and bases, salts, or other types of materials, for example, nanoparticles, cannot be described by this set of chemical property data. Multi-compartment models can also be set up for these other types of substances, but then the models have to be fed with property data that reflect the chemical and physical behavior of these substances (Praetorius et al., 2014).

Another fundamental assumption of multi-compartment models is that within each compartment, a chemical is always distributed homogeneously. Therefore, concentration gradients, for example, for a chemical in a lake or river, can only be described by several separate compartments, for example, for the layers of a stratified lake (epilimnion and hypolimnion) or different stretches of a river.

Depending on the problem investigated, multi-compartment environmental fate models can be built with lower or higher spatial and temporal resolution. Models with few compartments (e.g., air, water, soil, sediment) at steady state are generic and relatively coarse. They can be used to understand the basic features of a chemical's circulation in the multi-compartment system (Scheringer and MacLeod, 2021). But it is also possible to set up multi-compartment models with a high spatial resolution (many grid cells or spatial domains) (see MacLeod et al., 2011 (BETR-Global) and Glüge et al., 2016), and high temporal resolution (level IV models with short time steps) (Camenzuli et al., 2012). Of course, with increasing spatial and temporal resolution of the models, also more data on environmental conditions are needed.

Many models have been developed that operate with different amounts of required inputs and different adjustable settings and within different programming environments. Reviews of some of them are provided by Pistocchi et al. (2010) and Di Guardo et al. (2018). Some examples of applying these models to various settings from local to global include the investigation of polychlorinated biphenyls (PCBs) in an urban setting (Gasic et al., 2009), the global fate of the organochlorine insecticide endosulfan (Becker et al., 2011), and the global emissions and fate of perfluorooctanesulfonic acid (PFOS) (Wang et al., 2017).

Finally, for developers and users of any kind of environmental fate model, it is important to properly assess the validity of the model results. To help with this, Buser et al. (2012) developed a set of guidelines for good modeling practice.

6.4.3   Substance Evaluation in Terms of Persistence and Long-Range Transport Potential

Even without completing an effect assessment (discussed later in Sect. 6.5), a first, simplified evaluation of a substance can already be carried out on the basis of exposure using the two indicators of *persistence* and *long-range transport potential* (LRTP). These quantities describe the tendency of a substance to cause long-lasting and widespread exposures. Persistence serves as an indicator for the duration of an exposure and the LRTP as an indicator for an exposure's spatial extent. Even without precise knowledge of the effects of such exposures, substances with high persistence and LRTP can be considered to pose a threat to the environment (Scheringer and Berg, 1994). According to the precautionary principle, their use should therefore be avoided.

Chlorofluorocarbons (CFCs), for example, have a persistence of about 100 years due to their high chemical stability. According to their even global distribution, they also have a high spatial range of 40,000 km, the circumference of the Earth (the spatial range is a metric used to quantify the LRTP of chemicals (Scheringer, 2009)). Acid gases, such as NO_x and SO_x, can also be transported over distances of several hundred kilometers from, for example, Central Europe to Scandinavia. Both of these are examples of substantial spatial shifts of environmental exposure (and consequently effects) across regions.

According to the EU REACH legislation, a substance is considered persistent if its half-life exceeds a value of 40 days in freshwater or 120 days in soil (ECHA, 2017b). Similar criteria are used under the Stockholm Convention on Persistent Organic Pollutants. All these criteria are so-called single-media half-life criteria, and they are the established way of identifying persistent chemicals in chemical regulation schemes.

However, a chemical's persistence can also be measured by metrics other than these single-media half-life criteria. The *overall persistence* (P_{ov}) of a chemical is defined as a *residence time* or *turnover time* in a multi-compartment model, i.e., as the ratio of the total amount of chemical in the model system (the "stock"), M, and the flow of the chemical through the system, here denoted by E for the emission rate:

$$P_{ov} = \frac{M}{E} \tag{6.22}$$

The total amount of chemical in the system is the sum of the amounts in all individual compartments, i: $M = \sum_i m_i$, and the emission rate is, at steady state, equal to the total loss rate, which is given by the sum of all losses in the individual compartments: $E = \sum_i k_i m_i$ (with loss rate constants k_i). Accordingly:

$$P_{ov} = \frac{M}{E} = \frac{\sum_i m_i}{\sum_i k_i m_i} = \frac{1}{\sum_i k_i f_i} = \frac{1}{\bar{k}} \tag{6.23}$$

with $f_i = m_i/M$ being the fraction of the chemical's total amount that is present in compartment i and $\bar{k}$ denoting the weighted average of the first-order loss rate constants, k_i. Equation 6.23 shows that the overall persistence is the inverse of the weighted average of the loss rate constants in all media with the mass fractions, f_i, as weighting factors. This is a useful result because it shows how the overall persistence combines information about a chemical's degradability (the k_i) with information about the chemical's partitioning (the f_i). Many level III models calculate the P_{ov} of chemicals in this way.

However, there is also a drawback of P_{ov} as a metric of persistence. The f_i strongly depends on the way in which the chemical is emitted to the environment. Emission to soil, for example, generally leads to a higher fraction of chemical in soil than emission to air. In other words, the overall persistence includes an arbitrary element, namely the choice of the emission pathway and its influence on the f_i in Eq. 6.23.

A solution to this problem was proposed by Stroebe et al. (2004). If in a multi-compartment model a chemical is released to each compartment separately (i.e., 100% to air, 100% to water, etc.), a P_{ov} value is obtained for each of these model runs. The highest of all these P_{ov} values is a good estimate of the chemical's persistence in the temporal remote state (TRS). The persistence in the TRS describes how rapidly (or, rather, slowly) the most long-lived reservoir of the chemical in the multi-compartment model degrades. If a chemical's longest half-life is in soil, this most long-lived reservoir is in the soil. This reservoir consists of the amount of chemical that is still left when most of the chemical has been degraded in the other compartments. Therefore, the speed at which this last reservoir of the chemical disappears is a meaningful metric of the chemical's persistence, and notably, it is independent of the emission pathway (Stroebe et al., 2004).

This concept of calculating the greatest P_{ov} value in a model system and using it as an estimate of the chemical's persistence in the temporal remote state is also employed in the OECD P_{ov} and LRTP Screening Tool (Wegmann et al., 2009).

Persistence and LRTP can be determined from measurement data (laboratory measurements on degradation half-lives, measured concentrations of chemicals in the field) as well as from model calculations such as from the OECD's P_{ov} and LRTP Screening Tool. Substituting highly persistent and highly mobile substances with shorter-lived and less mobile substances can greatly limit environmental exposure (Scheringer, 1997, 2002).

6.5   Effect Assessment (Step 3)

In this next step of product risk assessment, the relationship between the level of exposure to a substance (*dose*) and the resulting effect (*response*) is investigated. This is called the *dose-response relationship* and can be estimated in a similar way for both humans and the environment.

There are two specific fields of toxicology that are used in understanding the dose-response relationship for a specific substance and effect. The field of *toxicokinetics* studies both the rate at which a substance is absorbed by an organism and the rates at which it is distributed, metabolized, and excreted. *Toxicodynamics* is the field that investigates the actual effect a substance then has on an organism, i.e., how the chemical interacts with the target site in the organism. Here, an effect refers to any measurable change in a biological parameter (known as an *endpoint*) that results from exposure and is dependent on the site of action, intensity, duration, and frequency of the exposure. During testing, changes in the test subject (often a cell or organism) are observed in comparison to a control group that has not been exposed to the chemical.

Table 6.1 provides some examples of this relationship between exposure, site of action, and resulting toxic effect. From the data forming the dose-response relationship, a *derived no-effect level* (DNEL, human health risk assessment) or a *predicted no-effect concentration* (PNEC, environmental risk assessment) is then derived by extrapolation. The DNEL or PNEC is considered appropriate for protecting human or environmental health.

The calculation of a DNEL or PNEC represents a classic line of thought in toxicology often characterized by the phrase "the dose makes the poison." However, it is important to keep in mind that there may be many chemicals with responses that do not necessarily increase with an increasing dose. Ongoing research has found that some substances can have higher effects at lower concentrations (Vandenberg et al., 2012), meaning that a DNEL is then not possible to define. Endocrine-disrupting chemicals (EDCs) that affect the functioning of the endocrine system are currently being closely reviewed for this reason and are discussed further in Sect. 6.5.2.

Table 6.1 Examples of toxic effects that could occur from acute vs. chronic exposure, listed according to where the chemical has an impact, known as the site of action

Exposure	Site of action	Example (compound)	Toxic effect
Acute	Local	Chlorine gas	Lung damage
	Systemic	Arsenic compounds	Hemolysis
	Mixed	Nitrogen oxides	Lung damage, methemoglobinemia
Chronic	Local	Sulfur dioxide	Bronchitis
	Systemic	Benzene	Leukemia
	Mixed	Tobacco smoke	Lung cancer, bladder cancer

Tools and methods that can be used to investigate the dose-response relationship include, for example, *quantitative structure-activity relationships* (QSARs), laboratory *in vitro* studies using cell cultures, *read-across* methods that apply results from existing studies on similar chemicals, experimental field studies, and, when necessary, laboratory *in vivo* studies using test animals.

During product development, the necessary depth of both the exposure and effect analyses can progressively increase based on the amount of data available for a chemical, its risk profile, and the tonnage being manufactured. For example, chemicals that are sold at amounts greater than 1000 t per year in the EU are required to be tested more extensively than those that are only sold at an amount of 10 t per year (ECHA, 2018).

In the interest of animal welfare, numerous standards have emerged to guide the appropriate testing methods to meet regulatory requirements, such as those developed by the Organisation for Economic Co-operation and Development (OECD) (OECD, 2019). Generally, *in vivo* testing should not be considered until all other relevant, available data and methods have been reviewed.

The next three subsections introduce the use of toxicokinetics (Sect. 6.5.1) and toxicodynamics (Sect. 6.5.2) in defining the dose-response relationship and then extrapolating these results to a no-effect level adequate for protecting humans and the environment (Sect. 6.5.3).

6.5.1   Toxicokinetics

Following exposure to a chemical, toxicokinetics studies the rate at which it is absorbed, distributed, metabolized, and excreted (ADME) by an organism. This describes the relation between the external and internal exposure to a chemical and, more specifically, defines the internal dose of a chemical as a function of space and time. In the human body, for example, a chemical can be taken up via the skin and gastrointestinal tract, as well as through the lungs. These are exposure routes each with a total surface area, barrier thickness, and resulting rate of perfusion (L/min) into the body (see Table 6.2).

Table 6.2 Typical surface area, barrier thickness, and rate of perfusion for each of the three main human exposure routes

Exposure route	Area	Thickness of barrier	Perfusion
Skin	$1.8\,\text{m}^2$	100–1000 µm	0.5 L/min
Gastrointestinal tract	$200\,\text{m}^2$	8–12 µm	1.5 L/min
Lung	$140\,\text{m}^2$	0.2–0.4 µm	6.0 L/min

Having crossed one of the barriers, a chemical can be distributed within the human body by either *passive transport* or *active transport* through cell membranes. Passive transport moves chemicals through cell membranes from areas of higher concentration to lower concentration via diffusion. In contrast, active transport can move chemicals from areas of lower concentration into areas of higher concentration by energy-dependent carriers (or pumps) within the cell membrane.

Once the chemical is distributed within the body, its fate is dependent on how the body metabolizes and excretes it. Nonvolatile chemicals are mainly metabolized and excreted through the liver and kidney. Enzymes can biotransform a chemical into water-soluble metabolites that can then be removed via urine.

Advances in the development of *physiologically based pharmacokinetic models* (PBPK models) have helped to better understand the toxicokinetics of chemicals in the body and estimate internal exposures and fates of metabolites (Karrer et al., 2018; Bachler et al., 2015; Brochot and Bois, 2005). Figure 6.15 shows a schematic of the flows and compartments of a PBPK model for the transfer of a chemical i from the lungs into the blood and then distributed between storage tissues j. Key parameters for the model include the size of the molecule, dissociation constant, octanol-water partition coefficient (K_{ow}), and metabolism rate constants. Notice how this use of compartments and flows between them is similar to the environmental fate models used in exposure assessment.

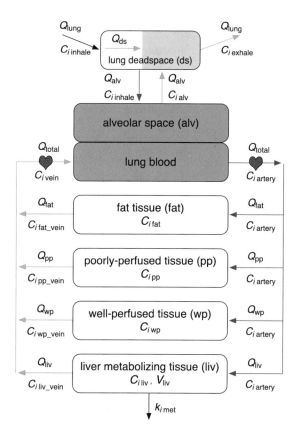

Fig. 6.15 Schematic of the flows and compartments involved in an example PBPK model investigating the transfer of a substance i from the lungs to the blood and then through distribution between storage tissues j in the body. Q_j, flow of blood/air through tissue j [L/min]; $k_{i,met}$, first-order rate constant of metabolism of substance i [min^{-1}]; C_{ij}, concentration of substance i in air, blood, or tissue j [mg/L]; V_{liv}, volume of liver tissue [L]

6.5.2   Toxicodynamics

The field of toxicodynamics investigates the cause of toxic effects of a chemical on an organism following exposure. A substance interacts with an organism at a *site of action* (e.g., a target tissue) and interacts with certain biochemical processes (known as the *mechanism*), which causes an adverse effect through a *mode of action* that results in damage to the organism at a *toxicological endpoint*. Table 6.3 provides an overview of various example sites of action and modes of toxic action through which a substance can cause an adverse effect.

Table 6.3 Examples of sites of action, mechanisms, and modes of action that describe how a chemical could cause a toxic effect (Escher and Hermens, 2002)

Site of action	Mechanism	Mode of action
Energy-transducing membranes	Ionophoric shuttle mechanisms	Uncoupling
	Blocking of quinone	Inhibition of the electron transport chain
	Blocking of proton channels	Inhibition of ATP synthesis/depletion of ATP
Photosynthetic membranes	Blocking of photosynthetic electron transport	Inhibition of photosynthesis
Proteins and peptides	Alkylation and oxidation	Damage and depletion of biomolecules
	Non-covalent and covalent binding to enzymes and receptors	Inhibition or competition, e.g., acetylcholine esterase, estrogen receptor
DNA or RNA	Base modification and damage	Mutagenicity

Mutagenesis and carcinogenesis are two modes of action commonly investigated during toxicity testing. *Mutagenesis* is the reaction of electrophilic substances with cell structures that results in the change of genetic information. This can cause a variety of toxic effects including cancers. *Carcinogenesis* is, specifically, the formation of cancers. Substances that cause carcinogenesis are known as *carcinogens* and can be either direct *initiators* that alter the cell structure and cause cancer or *promoters* that drive proliferation of the cancer but do not alter the cell themselves.

The *Ames test* is a common bioassay performed to identify the ability of a substance to cause mutations. It uses already mutated strains of *Salmonella* bacteria that are unable to produce an essential amino acid called histidine (and are therefore not able to multiply). These histidine-negative (His$^-$) bacteria are plated onto a set of Petri dishes containing combinations of a His$^-$ growth medium, the substance to be tested, and rat liver enzyme. The liver enzyme simulates the conditions in the human body where enzymes exist that could metabolize the compound and produce metabolites (which may themselves be mutagenic). When significant numbers of bacterial colonies are found to grow in the presence of the test substance, this shows that the substance had a mutagenic effect on the bacteria and was able to change them from His$^-$ to His$^+$.

Another mode of action that has been heavily discussed over the past decade is effects via the endocrine system. *Endocrine-disrupting chemicals* (EDCs) are chemicals that interfere with the body's natural endocrine system, which regulates the production of hormones. One or more of the many functions in the body controlled by hormones could be adversely affected by EDCs. Research on EDCs has challenged the classical assumption of "the dose makes the poison," which means that higher doses generally lead to stronger effects, by identifying substances that do not follow this type of relationship. Instead, some EDCs might cause stronger (or other) effects through exposure at lower concentrations than at higher concentrations.

EDCs can act either by (1) binding to a hormone receptor and causing hormonal action (*agonism*); (2) by binding to a receptor and, thereby, blocking a natural hormone and preventing action (*antagonism*); or by (3) interfering with important enzymes involved in hormone function. Various methods exist to test or screen for EDCs, with a common example being a test for interaction with receptors for the estrogen hormone. Screening methods also exist to use a chemical's structural data to predict its endocrine-disrupting potential by comparing it to the structure of natural hormones. Figure 6.16 depicts the structure of the natural female sex hormone estradiol and the industrial chemical nonylphenol used in the manufacturing of oil additives, antioxidants, and surfactants. Nonylphenol has been recognized as an EDC through interaction with estrogen receptors, and it is linked to causing feminization of aquatic organisms and a decrease of male fertility (ECHA, 2012).

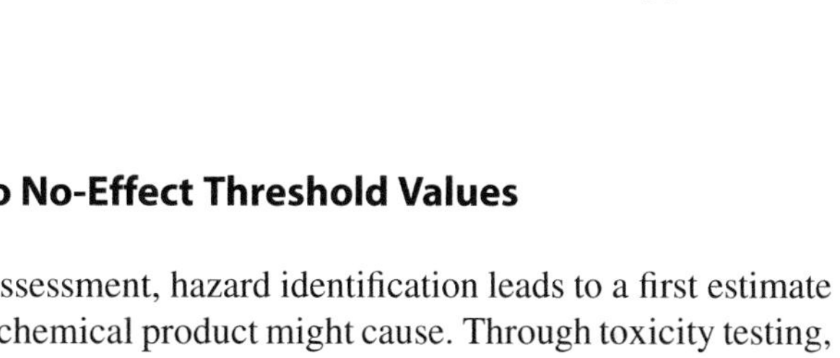

Fig. 6.16 The chemical structures of the female sex hormone estradiol and one of the nonylphenol isomers

6.5.3   Extrapolation to No-Effect Threshold Values

As the first step in the risk assessment, hazard identification leads to a first estimate of the types of effects that a chemical product might cause. Through toxicity testing, the dose-response relationship can be quantified, and from this threshold, values may be derived at which no adverse effects are expected to occur (for chemicals exhibiting a monotonous dose-response relationship). As shown in Fig. 6.17, the toxicity testing requirements in the EU increase with the annual production volume of the substance.

The dose-response relationship can be represented quantitatively in a variety of ways as shown in Fig. 6.18, including a frequency distribution, a cumulative distribution, or a linearized cumulative distribution. The linearized representation

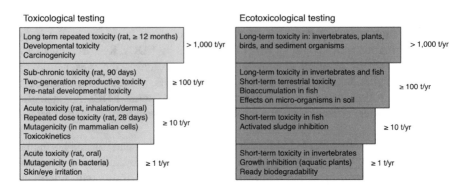

Fig. 6.17 A simplified overview of some of the toxicity testing requirements in the EU for industrial chemicals registered under REACH, depending on annual production volume (metric tons). Testing requirements in each volume step are in addition to the requirements in the step(s) below it. See REACH Annexes VII–X for full testing requirements (European Union, 2006)

of the dose-response relationship facilitates extrapolation from the quantified effect on the selected test organism to a maximum concentration deemed fit to protect humans or entire environmental ecosystems.

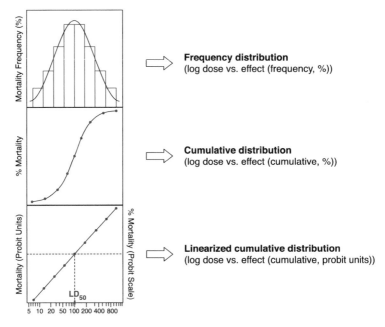

Fig. 6.18 Formats for statistical representation of a dose-response relationship: frequency distribution, cumulative distribution, and linearized cumulative distribution

Defining this safe level for humans can be difficult, since testing directly on humans is usually not possible or not allowed for ethical reasons. For environmental toxicity, testing is less restricted, but it has to be able to lead to a result that takes into account the thousands of different species that exist in an ecosystem, which can each have varying (and often unknown) sensitivities to the same substance.

To address these limitations, *assessment factors* are applied as a method to consider the uncertainty and variability within and between different species and to provide a greater level of protection. An estimated no-effect concentration for either humans or the environment can generally be calculated by applying an assessment factor as follows:

$$\text{No-Effect Concentration} = \frac{\text{Resulting effect level (from experiment)}}{\text{Assessment factor (AF)}} \quad (6.24)$$

The greater the level of knowledge about the dose and effect, the smaller the assessment factor that needs to be used. During an environmental analysis, for example, if only acute toxicity data are available on a substance for a single fish species, then a larger assessment factor can be chosen to offset this uncertainty and provide protection for other species as well. However, when studies on chronic toxicity in fish, daphnia (water fleas), and algae are all available for a substance, i.e., across three trophic levels, a lower assessment factor may be applied to calculate the no-effect concentration.

Different regulatory agencies recommend different approaches and guidelines for using assessment factors, and discussions are ongoing regarding best practices (Konietzka et al., 2014). However, an assessment factor is often applied to consider the uncertainty for each level of extrapolation needed. For example, it is customary that an additional extrapolation factor of 10 is included for each of the following extrapolations (Konietzka et al., 2014):

- The extrapolation of acute and sub-chronic effects to longer-term, chronic effects
- Interspecies extrapolation (e.g., from rats to humans)
- Intraspecies extrapolation (e.g., from an average group of humans to groups that have increased sensitivity, such as infants or the elderly)

Additional assessment factors may also be added to reflect qualitative aspects of the toxicity studies. This could include, for example, assessment factors to consider:

- A low level of data quality associated with the toxicological studies
- Increased severity of an effect that was observed or for endpoints deemed especially critical
- Uncertainty regarding potential properties of the substance that cannot be quantified (e.g., carcinogenicity)

All assessment factors to be applied for each of these aspects are then multiplied together to produce the total assessment factor used in Eq. 6.24. Other, more flexible approaches for the determination of assessment factors exist that make use of expert input and allow for the use of lower factors (ECETOC, 2003). These could be most relevant, for example, for assessments at the workplace, where control options and protection against exposure are better and particularly sensitive individuals do not need to be considered.

6.5.3.1 Human Toxicity

With results from toxicity studies, various no-effect threshold values can be calculated, including the *derived no-effect level (DNEL)*.

To determine the value of the *DNEL*, results from animal toxicity tests have to be extrapolated to humans, or results based on total population studies of humans need to be used (often including particularly vulnerable individuals). Following the structure of Eq. 6.24, the formula used is:

$$DNEL = \frac{LOAEL \text{ (or } NOAEL)}{AF} \tag{6.25}$$

- *DNEL* (Derived No-Effect Level): Highest concentration or dose of a substance that is not expected to cause any detectable adverse effect in humans
- *LOAEL* (Lowest Observed Adverse Effect Level): Lowest concentration or dose (experimentally or by monitoring) of a substance that causes an adverse effect in the studied organism
- *NOAEL* (No Observed Adverse Effect Level): Highest concentration or dose (experimentally or by monitoring) of a substance that does not cause any detectable adverse effect in the studied organism
- *AF*: Assessment factor

Calculated in a similar way, other existing human toxicity threshold values include:

- *Acceptable daily intake (ADI)*: Tolerable dose of a substance that can be taken in daily over a lifetime without damage to health [mg kg_{bw}^{-1} day^{-1}].
- *Arbeitsplatzgrenzwert (AGW)*: Translated into English as the "occupational exposure limit," these values were developed by the German government to define the average concentration in the air at the workplace to which acute or chronic adverse effects on employees are generally not to be expected. The values are defined according to a given period of exposure time, usually for an 8-hr daily exposure for 5 days a week and over an employee's entire working career (Beratungsgesellschaft für Arbeits- und Gesundheitsschutz, 2019).

6.5.3.2 Environmental Toxicity

For environmental toxicity, toxicity testing results can be extrapolated to a threshold value called the *predicted no-effect concentration (PNEC)* using an assessment factor. Just as for human toxicity, the method of assessment factors is widely used in practice to determine a PNEC given a level of uncertainty regarding an effect:

$$PNEC = \frac{LOEC \ (or \ NOEC)}{AF} \tag{6.26}$$

- *PNEC* (predicted no-effect concentration): the concentration below which exposure to a substance will likely not have a toxic effect
- *LOEC* (lowest observed effect concentration): the lowest concentration of a substance in a series of test concentrations that causes a statistically detectable toxic effect on the studied organism
- *NOEC* (no observed effect concentration): the highest concentration of a substance in a series of test concentrations that does not show a toxic effect on the studied organism
- *AF*: assessment factor

In addition to extrapolation to a *PNEC* value, another threshold value used in practice for the environment is the *maximum immission concentration (MIC)*. An MIC value refers to the concentration of a substance in air, water, and soil below which no damages to animals, plants, and humans are known to occur (in units of, for example, $\mu g/m^3$). The focus of the MIC is on preventing the long-term effects of chronic toxicity.

Within an effect assessment, not only direct but also indirect toxic effects from the accumulation of substances in the food chain can be considered.

6.5.3.3 General Aspects to Consider
In the course of the effect assessment, it is important to keep in mind some overarching aspects that can influence the assessment:

- *Inadequate characterization of hazard:* All available references and previous studies regarding the effects, exposure pathways, and exposed organisms should be considered.
- *Well-defined evaluation endpoints:* The determination of concentrations and doses that do not cause adverse effects in test organisms depends on how adverse effects are defined and on how they are investigated. In particular, the values of the *LOEC* and *NOEC* depend on the test concentrations chosen. A different selection of test concentrations generally leads to different *LOEC* and *NOEC* values (Laskowski, 1995).
- *Consider various methods:* Scientists often use and develop a wide range of methods to evaluate effects. While data accuracy and method transparency are important to ensure, it is advantageous to not limit assessments to using only a small set of previously established methods.
- *Diversity, variability, and temporal dynamics of ecosystems:* The resulting effects determined during testing have to be extrapolated to the real situation in the environment or in humans, and this is of course associated with uncertainties.

Keep in mind that there is a possibility that extremely sensitive species or persons are not protected by the established no-effect levels.

- *Limited validity of laboratory tests on aquatic organisms:* An initial estimate of environmental effects is often based on data for aquatic organisms (fish, daphnia, algae). These findings have, for example, only limited significance for terrestrial ecosystems.
- *Variability of product quality:* Especially in the early development stages of a product, the variability of product quality can complicate a prospective risk assessment.
- *Not all effects covered:* The concept of a safe effect threshold cannot be applied to all effects. For genotoxic carcinogens, for example, a tolerable risk level has to be established on the basis of the results from *in vivo* experiments.
- *Low-dose effects:* Remember that not all chemicals may follow the traditional thinking of "the dose makes the poison." Research is ongoing to understand substances such as some EDCs that may have a wide range of adverse effects that are greater at lower exposure levels.

6.6   Risk Characterization and Classification (Step 4)

In the next step of risk assessment, *risk characterization* compares the findings of the exposure and effect assessments in order to assess the vulnerability of a population or of an environmental compartment to the use of the substance. This is often done deterministically in the form of a risk quotient, but it can also be done probabilistically by using exposure and effect distributions. *Risk classification* then aims to determine whether the characterized risk is acceptable or not. This assessment step depends on set societal protection goals, but it still needs to be conducted in a transparent manner and in agreement with established criteria. As will be discussed more in this chapter and further in Chap. 9, a "subjective" classification of the risk does not mean that it is "arbitrary." If a risk is deemed not to be acceptable, subsequent *risk management* actions will then need to be identified and implemented in the next and final step of risk assessment (Sect. 6.7).

This section introduces a few of the fundamental calculation methods for assessing risks using both deterministic and probabilistic approaches. For risks to humans and the environment, a deterministic approach makes use of the *margin of safety (MOS)* and the *risk quotient (RQ)*, respectively. The basic methods to take into account the toxicity of chemical mixtures are also introduced.

6.6.1   Deterministic Human Health Risk: Margin of Safety (MOS)

Assessing the risk posed to humans aims to consider the risks for all identified critical effects across all population groups and relevant exposure routes. A deterministic description of this risk can be created by forming the quotient of the extrapolated safety threshold for adverse effects (e.g., derived no-effect level in humans (*DNEL*)) and the exposure level (e.g., predicted daily intake (PDI)). This quotient is defined as the *margin of safety (MOS)*:

$$MOS \ = \ \frac{DNEL}{PDI} \tag{6.27}$$

The MOS denotes the safety factor by which the current exposures fall below the threshold of effect, and its value should therefore be as high as possible. It can also be calculated as the ratio of other threshold limit values to exposure concentrations, for example, as the ratio of acceptable daily intake (ADI) to predicted daily intake (PDI). The *classification* of the risk depends on the set societal protection goals; however, a distinction is primarily made between two cases based on the resulting *MOS*:

(i) $MOS \ = \ \frac{DNEL}{PDI} \ > 1 \ \Rightarrow$ no action required
(ii) $MOS \ = \ \frac{DNEL}{PDI} \ < 1 \ \Rightarrow$ action required

If the $MOS < 1$, there is a need for action. Potentially the value of the MOS can be increased by developing more toxicity data, which then leads to the application of smaller assessment factors in the calculation of the $DNEL$. Otherwise, risk management measures need to be taken to reduce the risk (see Sect. 6.7).

If a $DNEL$ could not be determined during the effect assessment (e.g., for carcinogenic or sensitizing substances), a probabilistic approach (Sect. 6.6.3) can be used based on statistical or qualitative information on exposure and effects to describe the risk.

In addition to calculating an MOS as described here, another option for risk characterization is to compare the obtained $LOAEL$ or $NOAEL$ from animal testing directly (without an assessment factor) to the exposure. The resulting, larger MOS value can then be interpreted in consideration of the uncertainty by an expert assessor. This method, however, relies significantly on the assessor's background, experience, and judgment. In a regulatory situation, this approach would certainly need to be justified and may not be accepted by regulatory agencies.

6.6.2   Deterministic Environmental Risk: Risk Quotient (RQ)

Similar to a margin of safety for human health risk, risks posed to the environment are characterized through the calculation of a *risk quotient (RQ)*. This is defined as a quotient of the maximum predicted environmental concentration (*PEC*) and an extrapolated predicted no-effect concentration (*PNEC*):

$$RQ = \frac{PEC}{PNEC} \tag{6.28}$$

Since adverse environmental effects are often studied for a particular environmental compartment or species of test animal, several different risk quotients can be calculated, e.g., for surface water, groundwater, different soil types, worm- and fish-eating birds and mammals, microorganisms in sewage treatment plants, etc.

Just as for human health, the classification of the risk depends on the set protection goals; however, a distinction is made between the following two cases:

(i) $RQ = \frac{PEC}{PNEC} < 1 \Rightarrow$ no action required
(ii) $RQ = \frac{PEC}{PNEC} > 1 \Rightarrow$ action required

Improving the data quality here could also reduce the assessment factor used in calculating the *PNEC* and therefore reduce the risk quotient. Otherwise, risk management measures need to be taken to reduce the risk (see Sect. 6.7).

6.6.2.1 Critical Questions to Consider

Some important points to consider in characterizing a risk for both humans and the environment using a deterministic approach include:

- *Inadequate hazard detection:* Have effects, exposure pathways, and exposed populations been identified correctly?
- *Absence or uncertainty of data:* Have uncertainties been clearly addressed and any assumptions been checked? Is a sensitivity analysis needed? If the substance is highly hydrophobic, special care should be taken in reviewing the bioaccumulation and toxicity tests, as they can be inaccurate (see Stieger et al., 2014; Jonker and van der Heijden, 2007 and Stibany et al., 2020).
- *Variation of national and regional conditions:* Environmental exposure is influenced by factors such as geographic and climatic conditions, dietary habits, etc. Has the applicability of the results of the risk assessment to areas with different conditions been carefully reviewed?
- *Simultaneous chemical stressors:* For chemical stressors that act simultaneously, the combination of the effects needs to be considered (see Sect. 6.6.4).

6.6.3   Probabilistic Characterization of Risk

Deterministic approaches to characterizing a risk provide a simple, useful way forward when limited data exist for describing exposure and effect. However, in reality, both exposures and effects have a distribution of values that vary spatially, temporally, and within individual species. A commonly applied tool within probabilistic risk assessment is *Monte Carlo analysis*. Monte Carlo analysis uses a set of random inputs within a distribution that has realistic limits. It then models the system in order to generate a distribution of the outcomes.

To use this approach for chemical risks, a data-rich situation is required where the distributions of both exposure levels and effect thresholds are characterized by a sufficient number of empirical data points. The effect data need to represent relevant species from the environmental compartment under review. If these data are not available and the distributions are instead derived from inadequate data or general assumptions, the probabilistic approach may create a false impression of scientific rigor and accuracy. There are a few previous studies that serve as good examples of applying probabilistic approaches to the risk assessment of chemicals in the environment. These include a publication by Scheringer et al. (2002) assessing the pesticide methyl parathion, a study by Klaine et al. (1996) assessing the biocide dibromonitrilopropionamide (DBNPA), and another by Solomon et al. (1996) assessing the herbicide atrazine.

6.6.4   Mixture Toxicity

Both humans and the environment are exposed to numerous different substances simultaneously. However, chemical risk assessment has classically used a chemical-by-chemical approach focused on investigating the effects caused by each substance individually. Much less is known about the possible overall effects that exposure to several different, toxic substances at once might have. This combination of toxic effects is known as *mixture toxicity*. Exposure to multiple chemicals in a mixture can generally result in three different types of changes to the resulting overall toxic effect. Consider a mixture with the two chemicals A and B. Each chemical has its own, individual toxic effect, and when mixed together, they can interact in three different ways to influence the overall toxic effect resulting from exposure:

- *Additive*: overall effect = effect of A + effect of B
- *Synergistic*: overall effect > effect of A + effect of B
- *Antagonistic*: overall effect < effect of A + effect of B

Of these three, synergistic effects are clearly very important to carefully consider. The risk posed by chemical mixtures is an active area of research within the fields of toxicology and risk assessment (Rotter et al., 2018; Bornehag et al., 2019).

The overall effects caused by a mixture of chemicals that each have a similar mode and site of action can be quantified using *concentration addition* (Backhaus and Faust, 2012). In concentration addition, the risk quotient of the mixture is the sum of the risk quotients of the individual mixture components and defined as:

$$RQ_{\text{mix}} = \sum_{i=1}^{n} \frac{c_i}{ECx_i} \qquad (6.29)$$

- RQ_{mix}: risk quotient of the mixture containing n chemicals
- c_i: concentration of chemical i within the mixture
- ECx_i: effect concentration of chemical i for an effect level of $x\%$ of the test organisms affected

Concentration addition can therefore be applied, for example, for compounds that have many congeners such as polychlorinated dibenzodioxins and dibenzofurans (PCDD/Fs). Each congener acts in the same way and at the same site of action, but each has a different toxic potency. *Toxic equivalency factors (TEFs)* can be defined to describe each congener's potency as a fraction of the most potent congener. The total concentration of congeners in a mixture can then be defined in terms of toxic equivalents of the most potent congener:

$$c_{\text{total}} = \sum_{i=1}^{n} c_i \times TEF_i \qquad (6.30)$$

- c_{total}: total concentration of the mixture in terms of equivalents of the most potent congener
- c_i: concentration of chemical i in the mixture
- TEF_i: toxic equivalency factor of chemical i

For example, for three chemicals with a similar mode and site of action, the mixture toxicity of acetone, 1-butanol, and chloroform can be calculated under the assumption of concentration addition. For each substance, Table 6.4 provides example values of measured workplace concentrations along with occupational exposure limits (OELs) set within EU regulations. The total risk quotient of the mixture (RQ_{mix}) can be calculated as:

$$RQ_{\text{mix}} = \sum_i \frac{c_i}{OEL} = \frac{375}{500} + \frac{20}{200} + \frac{1}{2} = 1.35 \qquad (6.31)$$

Individually, these chemicals would each have a risk quotient below one and be seen as no cause for concern. However, considering their combined effects through concentration addition, their mixture results in a risk quotient greater than one and in a need for risk reduction.

Table 6.4 Example of measured air concentrations at a manufacturing site and occupational exposure limits (OELs) for an exposure time of 8 hr in the European Union for three chemicals (European Union, 2000)

Substance	Workplace concentration [ppm]	Occupational exposure limit (OEL) [ppm]
Acetone	375	500
Butanone	20	200
Chloroform	1	2

Effects from a mixture of chemicals with dissimilar modes and sites of action can be quantified through using *effect addition* (also known as *independent action*). In such mixtures, the chemicals exert their different toxic effects independently of each other, and calculating their overall effect requires information about the effect caused by each chemical within the mixture at exactly the concentration at which it is present in the mixture. In this case, the overall effect (E) of the mixture can be derived as:

$$E(c_{mix}) = E(c_1 + \ldots + c_n) = 1 - \prod_{i=1}^{n}(1 - E(c_i)) \qquad (6.32)$$

- E: overall effect (scaled to the range 0–1)
- c_{mix}: total concentration of the mixture
- c_i: concentration of chemical i
- $E(c_i)$: effect of concentration of chemical i (scaled to the range 0–1)

Effect addition is a more challenging concept to quantitatively calculate, and the literature provides further explanation and examples (Backhaus and Faust, 2012; Backhaus et al., 2000; Altenburger et al., 2000).

6.7 Risk Management (Step 5)

If a risk posed by a substance has been characterized and then classified as not acceptable, the next step of *risk management* is needed. This step considers the range of actions or measures available to reduce the current risk and then selects and applies the best one taking into account set goals and limitations. Often, a *risk-benefit analysis* is used to compare different options, and discussions surrounding the decision heavily involve both science and policy. This can sometimes require a dialogue with external stakeholders when sensitive topics or risks with high impacts are involved, and this is discussed further in Chap. 9.

Risk management options for a chemical product can be introduced at different stages in the design process and tailored to effectively target reduction either of the hazard or of the exposure. For example, risk management measures focused on reducing hazards can include:

- Investigation during the development of the molecular structure to avoid inherently hazardous properties; this includes, for example, 2 of the 12 principles of Green Chemistry: designing safer chemicals and designing less persistent chemicals that can easily degrade (Anastas and Warner, 1998).
- Improvement of the product formulation, for example, by aiming to use easily degradable excipients.
- Elimination of critical impurities in the product formulation.

Risk management focused on reducing exposure can include:

- Improving efficiency of the use or application process and, thereby, reducing exposure

- Having a suitable commercial form and packaging of the product
- Developing and enforcing product recycling and disposal regulations
- Providing comprehensive product information and training to users (e.g., personal protection devices as well as control and emergency measures)
- From a long-term perspective, initiating the search for alternative products according to the concept of Green Chemistry

After implementing any risk reduction measures, a final, recurring aspect of risk management is to set up a plan to *monitor and review* the actions taken. This should include a set process over a defined space and time frame that checks the effectiveness of the risk reduction measures and alerts to any new adverse impacts on human health and the environment in the future.

6.8   Example: Risk Assessment of a Reactive Cotton Dye

Finally, the five steps introduced for product risk assessment are applied here to an illustrative example of a cotton dye.

Problem Description: A reactive dye of the triazine chemical class (structure shown in Fig. 6.19) has been developed and needs to be risk assessed before it can be registered and placed on the market. The following two questions need to be answered: (1) To what extent does long-term exposure from using the dye in a cotton-processing factory pose health risks to the employees? (2) Does the dye pose a risk to the local ecosystem through release into the environment following wastewater treatment?

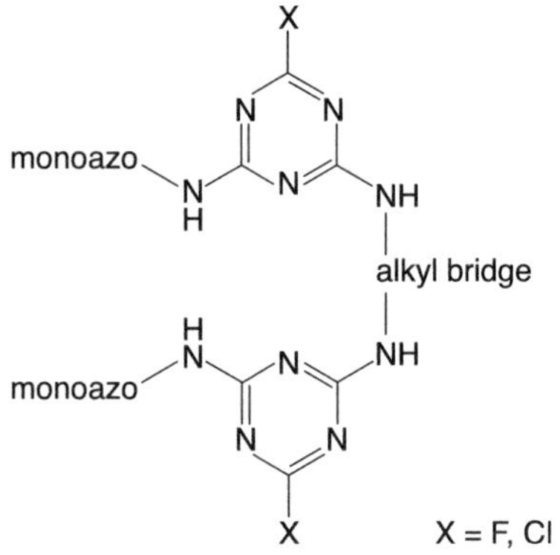

Fig. 6.19 Chemical structure of the reactive dye investigated. The monoazo group contains the chromophore, which defines the color of the dye

In each of the following subsections, information on the dye's life cycle, chemical properties, and toxicity are provided. This information should be used to provide the factory with appropriate recommendations in answering these two questions. Data and relevant calculations are presented separately for assessing the risk to human health and the environment. The values provided here are based on typical values for such reactive triazine dyes.

6.8.1 Hazard Identification (Step 1)

Through a review of existing toxicity data on this particular dye and on chemicals within the same class, it was identified that:

- In preparing the dye liquor starting from powder, the sensitizing effect and chronic toxicity of the dye can be a potential hazard to the health of the employees.
- The dye may be toxic to the aquatic environment and may not be adequately removed during wastewater treatment processes.

6.8.2 Exposure Assessment (Step 2)

Investigation of potential human exposure pathways results in the following observations about the dye:

- Used in open systems for dyeing and printing cotton.
- Used by employees in the form of a nonvolatile powder.
- No bioaccumulation is expected since log $K_{ow} \cong -4$.
- Reported average concentrations of dye dusts in dyeing operations from an earlier monitoring study at a similar factory are 0.1–$0.3 \, mg/m^3$ air.
- Average air volume inhaled by the employees during an 8-hr shift is $10 \, m^3$.

From this information, the predicted daily intake (PDI) via inhalation can be calculated (with an assumed average body weight of 70 kg):

$$PDI = \frac{0.3 \, mg/m^3 \times 10 \, m^3/d}{70 \, kg} = 0.043 \, mg \, kg_{bw}^{-1} \, d^{-1}$$

Investigation of potential environmental exposure pathways results in the following information:

- Mass of dyed material per day (textiles) $(W_1) = 3000 \, kg/d$
- Dye used per amount of dyed material $(W_2) = 10 \, g/kg$
- Fixation level of dye onto the cotton $(F) = 85\%$
- Amount of wastewater produced $(Q) = 300 \, L/kg$ textiles
- Fraction of dye removed from the wastewater during treatment $(P) = 10\%$
- Dilution factor of the treated water upon release $(V) = $ approximately 100

The predicted environmental concentration from the release of the dye following wastewater treatment (PEC_{local}) can then be calculated through Eq. 6.17. This models the emissions as a point source:

$$PEC_{local} = \frac{W_1 \times W_2 \times \frac{100-F}{100} \times \frac{100-P}{100}}{Q \times W_1 \times V}$$

$$= \frac{3000\,\text{kg/d} \times 10\,\text{g/kg} \times \frac{100-85}{100} \times \frac{100-10}{100}}{(300\,\text{L/kg} \times 3000\,\text{kg/d}) \times 100} = 0.045\,\text{mg/L}$$

6.8.3   Effect Assessment (Step 3)

Toxicity testing data for triazine dye structures indicate:

- Acute toxicity: LD_{50} (rat, oral) $>2000\,\text{mg/kg}$.
- No mutagenic effects.
- Substance is sensitizing to the skin in animal experiments.
- A 28-day dosing trial at $1000\,\text{mg kg}_{bw}^{-1}\,\text{d}^{-1}$ (rat, orally administered) resulted in minimal anemic effects. This value can be assumed to be lowest observed adverse effect level (LOAEL).

For the determination of the derived no-effect level (DNEL) on the basis of this 28-day study, an assessment factor (AF) of 10,000 could be taken into account. This is calculated by multiplying individual assessment factors of 10 for extrapolation from sub-chronic to chronic effects, from rats to humans, and from average humans to more sensitive groups, and considering that only one toxicity study is used. According to Eq. 6.25, this results in a DNEL of $0.1\,\text{mg kg}_{bw}^{-1}\,\text{d}^{-1}$:

$$DNEL = \frac{LOAEL}{AF} = \frac{1000\,\text{mg kg}_{bw}^{-1}\,\text{d}^{-1}}{10,000} = 0.1\,\text{mg kg}_{bw}^{-1}\,\text{d}^{-1}$$

Aquatic toxicity testing of the dye provides multiple results for different organisms:

- Fish toxicity: $LC_{50} >250\,\text{mg/L}$ (96-hr test with Zebrafish)
- Daphnia toxicity: $EC_{50} >250$ mg/L (48-hr test)
- Algae toxicity: $EC_{50} >100\,\text{mg/L}$ (72-hr test)

An assessment factor of 1000 can be applied for extrapolation from acute to chronic effects and from a single species to many aquatic species and that only one toxicity study is used. The predicted no-effect concentration for the aquatic environment ($PNEC_{aquatic}$) can then be calculated as (Eq. 6.26):

$$PNEC_{aquatic} = \frac{\text{lowest acute toxicity value}}{AF} = \frac{100\,\text{mg/L}}{1000} = 0.1\,\text{mg/L}$$

6.8.4   Risk Characterization and Classification (Step 4)

Given the calculated exposure and effect values, a deterministic risk quotient can be used to characterize the risks of dye use to human health and the environment.

For human toxicity, the dust inhalation PDI represents a chronic toxicity, and its risk can be characterized using the margin of safety (MOS) approach from Eq. 6.27:

$$MOS = \frac{DNEL}{PDI} = \frac{0.1 \text{ mg kg}_{bw}^{-1} \text{ d}^{-1}}{0.043 \text{ mg kg}_{bw}^{-1} \text{ d}^{-1}} = 2.3$$

As the calculated MOS value is greater than 1, no harm to the employees is expected to occur. However, to formally classify this risk according to any set protection goals, this result should be interpreted considering that:

- The calculated MOS value, relative to the uncertainty, is still close to 1.
- The uncertainty in this case is largely due to the extrapolation of the test results from rats to humans.
- The assumed exposures are based on measurements and are expected to be within the correct order of magnitude.
- It was assumed that all dust is respirable, although it may not be.
- The chosen extrapolation factor of 10,000 is very conservative, and it is expected that the no-effect level is in reality likely higher.

Another aspect to consider is that the development of sensitization by skin contact depends on the level of exposure. Employees who have already been sensitized could show reactions even at very low exposure levels, and a characterization using a risk quotient (such as MOS) cannot be used in this case. Instead, the risk characterization needs to be based on the statistically recorded occurrence of the sensitizing effect of similar substances and the likelihood of dermal contact, meaning that the occupational hygiene at the site needs to be more carefully evaluated.

The environmental risk can be characterized through the use of a risk quotient comparing the PEC following wastewater treatment to the PNEC (Eq. 6.28):

$$\text{Risk Quotient} = \frac{PEC}{PNEC} = \frac{0.045 \text{ mg/L}}{0.1 \text{ mg/L}} = 0.45$$

Since the risk quotient is less than 1, no harm to the aquatic ecosystem is expected to occur. However, in the interpretation of this result for the classification of the risk, the following should be considered:

- The risk quotient is still relatively close to 1.
- The outflow of wastewater can occur only intermittently, meaning that at times possibly higher concentrations occur than was calculated to be the average.

- The uncertainty is relatively high as many extrapolation steps had to be made.
- The assessment factor of 1000, however, is relatively conservative and may compensate for this uncertainty.

6.8.5   Risk Management (Step 5)

To reduce the risk of adverse effects on the employees, safety measures could be considered to reduce exposure to the dye, including:

- Using a liquid formulation of the dye or a formulation as low-dust granules
- Applying closed handling processes during dyeing
- Improving building ventilation
- Using personal protective equipment

To reduce the risk of adverse effects on the environment, risk reduction measures that could be considered include:

- Minimizing residues of the dye in the dyeing liquor, e.g., by maximum extraction
- Adsorption to activated carbon
- Concentrating and desalinating the wastewater using reverse osmosis

While these risk reduction options focus on minimizing exposure, an alternative dye could also be sought that has reduced inherent hazardous properties.

References

Altenburger R, Backhaus T, Boedeker W, Faust M, Scholze M, Grimme LH (2000) Predictability of the toxicity of multiple chemical mixtures to Vibrio fischeri: Mixtures composed of similarly acting chemicals. Environmental Toxicology and Chemistry 19(9):2341–2347, https://doi.org/10.1002/etc.5620190926, http://doi.wiley.com/10.1002/etc.5620190926

Anastas P, Warner J (1998) Green Chemistry: Theory and Practice. Oxford University Press

Arnot JA, Arnot M, Mackay D, Couillard Y, MacDonald D, Bonnell M, Doyle P (2010) Molecular size cutoff criteria for screening bioaccumulation potential: Fact or fiction? Integrated Environmental Assessment and Management 6(2):210–224, https://doi.org/10.1897/IEAM_2009-051.1, http://doi.wiley.com/10.1897/IEAM_2009-051.1

Bachler G, von Goetz N, Hungerbühler K (2015) Using physiologically based pharmacokinetic (PBPK) modeling for dietary risk assessment of titanium dioxide (TiO2) nanoparticles. Nanotoxicology 9(3):373–380, https://doi.org/10.3109/17435390.2014.940404

Backhaus T, Faust M (2012) Predictive Environmental Risk Assessment of Chemical Mixtures: A Conceptual Framework. Environmental Science & Technology 46(5):2564–2573, https://doi.org/10.1021/es2034125, https://pubs.acs.org/doi/10.1021/es2034125

Backhaus T, Altenburger R, Boedeker W, Faust M, Scholze M, Grimme LH (2000) Predictability of the toxicity of a multiple mixture of dissimilarly acting chemicals to Vibrio fischeri. Environmental Toxicology and Chemistry 19(9):2348–2356, https://doi.org/10.1002/etc.5620190927

Becker L, Scheringer M, Schenker U, Hungerbühler K (2011) Assessment of the environmental persistence and long-range transport of endosulfan. Environmental Pollution 159(6):1737–

1743, https://doi.org/10.1016/j.envpol.2011.02.012, http://linkinghub.elsevier.com/retrieve/pii/ S0269749111000868

Beratungsgesellschaft für Arbeits- und Gesundheitsschutz (2019) Arbeitsplatzgrenzwert (AGW) - Definition https://www.bfga.de/arbeitsschutz-lexikon-von-a-bis-z/fachbegriffe-a-b/ agw-fachbegriff/

Bornehag C, Kitraki E, Stamatakis A, Panagiotidou E, Rudén C, Shu H, Lindh C, Ruegg J, Gennings C (2019) A Novel Approach to Chemical Mixture Risk Assessment - Linking Data from Population-Based Epidemiology and Experimental Animal Tests. Risk Analysis 39(10):2259–2271, https://doi.org/10.1111/risa.13323, https://onlinelibrary.wiley.com/doi/abs/ 10.1111/risa.13323

Brochot C, Bois FY (2005) Use of a Chemical Probe to Increase Safety for Human Volunteers in Toxicokinetic Studies. Risk Analysis 25(6):1559–1571, https://doi.org/10.1111/j.1539-6924. 2005.00682.x, http://doi.wiley.com/10.1111/j.1539-6924.2005.00682.x

Buser AM, MacLeod M, Scheringer M, Mackay D, Bonnell M, Russell MH, DePinto JV, Hungerbühler K (2012) Good modeling practice guidelines for applying multimedia models in chemical assessments. Integrated Environmental Assessment and Management 8(4):703–708, https://doi.org/10.1002/ieam.1299, http://doi.wiley.com/10.1002/ieam.1299

Camenzuli L, Scheringer M, Gaus C, Ng CA, Hungerbühler K (2012) Describing the environmental fate of diuron in a tropical river catchment. Science of The Total Environment 440:178–185, https://doi.org/10.1016/j.scitotenv.2012.07.037, https://linkinghub.elsevier.com/ retrieve/pii/S0048969712009825

Di Guardo A, Gouin T, MacLeod M, Scheringer M (2018) Environmental fate and exposure models: advances and challenges in 21st century chemical risk assessment. Environmental Science: Processes & Impacts 20(1):58–71, https://doi.org/10.1039/C7EM00568G, http://xlink. rsc.org/?DOI=C7EM00568G

ECHA (2011) Guidance on information requirements and chemical safety assessment Part B: Hazard assessment. Tech. rep., https://echa.europa.eu/documents/10162/13643/information_ requirements_part_b_en.pdf/7e6bf845-e1a3-4518-8705-c64b17cecae8

ECHA (2012) Support Document for the identification 4-Nonylphenol, branched and linear as substances of very high concern. Tech. rep., https://echa.europa.eu/documents/10162/3024c102-20c9-4973-8f4e-7fc1dd361e7d

ECHA (2016) Guidance on information requirements and chemical safety assessment Chapter R.16: Environmental exposure estimation. Tech. rep., http://echa.europa.eu/doc/FINAL_MB_ 30_2007_Consultation_procedure_on_guidance.pdf

ECHA (2017a) Evaluation under REACH: Progress Report 2017. Tech. rep., https://echa.europa. eu/documents/10162/13628/evaluation_under_reach_progress_en.pdf/24c24728-2543-640c-204e-c61c36401048

ECHA (2017b) Guidance on Information Requirements and Chemical Safety Assessment. Chapter R.11: PBT/vPvB assessment. Tech. rep., European Chemicals Agency, https://doi.org/10.2823/ 128621, https://echa.europa.eu/documents/10162/13632/information_requirements_r11_en.pdf

ECHA (2018) Guidance on Information Requirements and Chemical Safety Assessment. https:// echa.europa.eu/guidance-documents/guidance-on-information-requirements-and-chemical-safety-assessment

ECHA (2019) Progress in evaluation in 2018. https://echa.europa.eu/dossier-evaluation-progress-2018

ECHA (2020) Registered substances. https://echa.europa.eu/information-on-chemicals/registered-substances

Escher BI, Hermens JLM (2002) Modes of Action in Ecotoxicology: Their Role in Body Burdens, Species Sensitivity, QSARs, and Mixture Effects. Environmental Science & Technology 36(20):4201–4217, https://doi.org/10.1021/es015848h, http://pubs.acs.org/doi/abs/10. 1021/es015848h

European Centre for Ecotoxicology and Toxicology of Chemicals (2003) Derivation of Assessment Factors for Human Health Risk Assessment - Technical Report 86. Tech. rep., http://www.ecetoc.org/publication/tr-086-derivation-of-assessment-factors-for-human-health-risk-assessment/

European Union (2000) Commission Directive 2000/39/EC of 8 June 2000 establishing a first list of indicative occupational exposure limit values in implementation of Council Directive 98/24/EC on the protection of the health and safety of workers from the risks related to chemi. https://eur-lex.europa.eu/legal-content/EN/TXT/?uri=CELEX:02000L0039-20180821

European Union (2006) Regulation (EC) No 1907/2006 of the European Parliament and of the Council concerning the Registration, Evaluation, Authorisation and Restriction of Chemicals (REACH). https://eur-lex.europa.eu/legal-content/en/TXT/?uri=CELEX:02006R1907-20180509

Frederiksen H, Skakkebaek NE, Andersson AM (2007) Metabolism of phthalates in humans. Molecular Nutrition & Food Research 51(7):899–911, https://doi.org/10.1002/mnfr.200600243, http://doi.wiley.com/10.1002/mnfr.200600243

Gasic B, Moeckel C, MacLeod M, Brunner J, Scheringer M, Jones KC, Hungerbühler K (2009) Measuring and Modeling Short-Term Variability of PCBs in Air and Characterization of Urban Source Strength in Zurich, Switzerland. Environmental Science & Technology 43(3):769–776, https://doi.org/10.1021/es8023435, http://pubs.acs.org/doi/abs/10.1021/es8023435

Glüge J, Bogdal C, Scheringer M, Hungerbühler K (2016) What determines PCB concentrations in soils in rural and urban areas? Insights from a multi-media fate model for Switzerland as a case study. Science of The Total Environment 550:1152–1162, https://doi.org/10.1016/j.scitotenv.2016.01.097, https://linkinghub.elsevier.com/retrieve/pii/S0048969716300912

Gomis MI, Vestergren R, MacLeod M, Mueller JF, Cousins IT (2017) Historical human exposure to perfluoroalkyl acids in the United States and Australia reconstructed from biomonitoring data using population-based pharmacokinetic modelling. Environment International 108(August):92–102, https://doi.org/10.1016/j.envint.2017.08.002, http://linkinghub.elsevier.com/retrieve/pii/S0160412017309601

Grimm FA, Hu D, Kania-Korwel I, Lehmler HJ, Ludewig G, Hornbuckle KC, Duffel MW, Bergman A, Robertson LW (2015) Metabolism and metabolites of polychlorinated biphenyls. Critical Reviews in Toxicology 45(3):245–272, https://doi.org/10.3109/10408444.2014.999365, https://www.tandfonline.com/doi/full/10.3109/10408444.2014.999365

Gyalpo T, Fritsche L, Bouwman H, Bornman R, Scheringer M, Hungerbühler K (2012) Estimation of human body concentrations of DDT from indoor residual spraying for malaria control. Environmental Pollution 169:235–241, https://doi.org/10.1016/j.envpol.2012.04.032, https://linkinghub.elsevier.com/retrieve/pii/S0269749112002266

Jonker MTO, van der Heijden SA (2007) Bioconcentration Factor Hydrophobicity Cutoff: An Artificial Phenomenon Reconstructed. Environmental Science & Technology 41(21):7363–7369, https://doi.org/10.1021/es0709977, https://pubs.acs.org/doi/10.1021/es0709977

Karrer C, Roiss T, von Goetz N, Gramec Skledar D, Peterlin Mašič L, Hungerbühler K (2018) Physiologically Based Pharmacokinetic (PBPK) Modeling of the Bisphenols BPA, BPS, BPF, and BPAF with New Experimental Metabolic Parameters: Comparing the Pharmacokinetic Behavior of BPA with Its Substitutes. Environmental Health Perspectives 126(7):077,002, https://doi.org/10.1289/EHP2739, https://ehp.niehs.nih.gov/doi/10.1289/EHP2739

Klaine SJ, Cobb GP, Dickerson RL, Dixon KR, Kendall RJ, Smith EE, Solomon KR (1996) An ecological risk assessment for the use of the biocide, dibromonitrilopropionamide (DBNPA), in industrial cooling systems. Environmental Toxicology and Chemistry 15(1):21–30, https://doi.org/10.1002/etc.5620150104, http://doi.wiley.com/10.1002/etc.5620150104

Koch HM, Angerer J (2007) Di-iso-nonylphthalate (DINP) metabolites in human urine after a single oral dose of deuterium-labelled DINP. International Journal of Hygiene and Environmental Health 210(1):9–19, https://doi.org/10.1016/j.ijheh.2006.11.008, https://linkinghub.elsevier.com/retrieve/pii/S1438463906001179

Koch HM, Christensen KLY, Harth V, Lorber M, Brüning T (2012) Di-n-butyl phthalate (DnBP) and diisobutyl phthalate (DiBP) metabolism in a human volunteer after single oral doses. Archives of Toxicology 86(12):1829–1839, https://doi.org/10.1007/s00204-012-0908-1, http://link.springer.com/10.1007/s00204-012-0908-1

Konietzka R, Schneider K, Ritter L (2014) Extrapolation Factors and Safety Factors in Toxicology. In: Regulatory Toxicology, Springer Berlin Heidelberg, Berlin, Heidelberg, pp 431–438, https://doi.org/10.1007/978-3-642-35374-1_59, http://link.springer.com/10.1007/978-3-642-35374-1_59

Laskowski R (1995) Some Good Reasons to Ban the Use of NOEC, LOEC and Related Concepts in Ecotoxicology. Oikos 73(1):140, https://doi.org/10.2307/3545738, https://www.jstor.org/stable/3545738?origin=crossref

van Leeuwen C, Vermeire T (2007) Risk Assessment of Chemicals. Springer Netherlands, Dordrecht, https://doi.org/10.1007/978-1-4020-6102-8, http://link.springer.com/10.1007/978-1-4020-6102-8

Mackay D (2001) Multimedia Environmental Models. CRC Press, https://doi.org/10.1201/9781420032543, https://www.taylorfrancis.com/books/9781420032543

MacLeod M, von Waldow H, Tay P, Armitage JM, Wöhrnschimmel H, Riley WJ, McKone TE, Hungerbühler K (2011) BETR global - A geographically-explicit global-scale multimedia contaminant fate model. Environmental Pollution 159(5):1442–1445, https://doi.org/10.1016/j.envpol.2011.01.038, https://linkinghub.elsevier.com/retrieve/pii/S0269749111000601

OECD (2012) Test No. 305: Bioaccumulation in Fish: Aqueous and Dietary Exposure. Tech. rep., https://doi.org/10.1787/9789264185296-en, https://www.oecd-ilibrary.org/environment/test-no-305-bioaccumulation-in-fish-aqueous-and-dietary-exposure_9789264185296-en

OECD (2019) OECD Test Guidelines for Chemicals. https://www.oecd.org/env/ehs/testing/oecdguidelinesforthetestingofchemicals.htm

Pistocchi A, Sarigiannis D, Vizcaino P (2010) Spatially explicit multimedia fate models for pollutants in Europe: State of the art and perspectives. Science of The Total Environment 408(18):3817–3830, https://doi.org/10.1016/j.scitotenv.2009.10.046, http://linkinghub.elsevier.com/retrieve/pii/S0048969709010274

Praetorius A, Tufenkji N, Goss KU, Scheringer M, von der Kammer F, Elimelech M (2014) The road to nowhere: equilibrium partition coefficients for nanoparticles. Environ Sci: Nano 1(4):317–323, https://doi.org/10.1039/C4EN00043A, http://xlink.rsc.org/?DOI=C4EN00043A

Ritter R, Scheringer M, MacLeod M, Schenker U, Hungerbühler K (2009) A Multi-Individual Pharmacokinetic Model Framework for Interpreting Time Trends of Persistent Chemicals in Human Populations: Application to a Postban Situation. Environmental Health Perspectives 117(8):1280–1286, https://doi.org/10.1289/ehp.0900648, https://ehp.niehs.nih.gov/doi/10.1289/ehp.0900648

Ritter R, Scheringer M, MacLeod M, Moeckel C, Jones KC, Hungerbühler K (2011) Intrinsic Human Elimination Half-Lives of Polychlorinated Biphenyls Derived from the Temporal Evolution of Cross-Sectional Biomonitoring Data from the United Kingdom. Environmental Health Perspectives 119(2):225–231, https://doi.org/10.1289/ehp.1002211, https://ehp.niehs.nih.gov/doi/10.1289/ehp.1002211

Rotter S, Beronius A, Boobis AR, Hanberg A, van Klaveren J, Luijten M, Machera K, Nikolopoulou D, van der Voet H, Zilliacus J, Solecki R (2018) Overview on legislation and scientific approaches for risk assessment of combined exposure to multiple chemicals: the potential EuroMix contribution. Critical Reviews in Toxicology 48(9):796–814, https://doi.org/10.1080/10408444.2018.1541964

Scheringer M (1997) Characterization of the Environmental Distribution Behavior of Organic Chemicals by Means of Persistence and Spatial Range. Environmental Science & Technology 31(10):2891–2897, https://doi.org/10.1021/es970090g, http://pubs.acs.org/doi/abs/10.1021/es970090g

Scheringer M (2002) Persistence and Spatial Range of Environmental Chemicals. Wiley-VCH Verlag GmbH & Co. KGaA, Weinheim, FRG, https://doi.org/10.1002/3527607463, http://doi.wiley.com/10.1002/3527607463

Scheringer M (2009) Long-range transport of organic chemicals in the environment. Environmental Toxicology and Chemistry 28(4):677, https://doi.org/10.1897/08-324R.1, http://doi.wiley.com/10.1897/08-324R.1

Scheringer M, Berg M (1994) Spatial and temporal range as measures of environmental threat. Fresenius Environmental Bulletin 4(8):493–498

Scheringer M, MacLeod M (2021) The Small World and Small Region Models – Multimedia Environmental Fate Models for Application in Teaching. https://zenodo.org/record/4438314

Scheringer M, Steinbach D, Escher B, Hungerbühler K (2002) Probabilistic approaches in the effect assessment of toxic chemicals. Environmental Science and Pollution Research 9(5):307–314, https://doi.org/10.1007/BF02987572, http://link.springer.com/10.1007/BF02987572

Schwarzenbach RP, Gschwend PM, Imboden DM (2016) Environmental Organic Chemistry, 3rd edn. John Wiley & Sons, Inc., Hoboken, NJ, USA, https://www.wiley.com/en-bw/Environmental+Organic+Chemistry%2C+3rd+Edition-p-9781118767047

Solomon KR, Baker DB, Richards RP, Dixon KR, Klaine SJ, La Point TW, Kendall RJ, Weisskopf CP, Giddings JM, Giesy JP, Hall LW, Williams WM (1996) Ecological risk assessment of atrazine in North American surface waters. Environmental Toxicology and Chemistry 15(1):31–76, https://doi.org/10.1002/etc.5620150105, http://doi.wiley.com/10.1002/etc.5620150105

Stibany F, Schmidt SN, Mayer P, Schäffer A (2020) Toxicity of dodecylbenzene to algae, crustacean, and fish - Passive dosing of highly hydrophobic liquids at the solubility limit. Chemosphere 251:126,396, https://doi.org/10.1016/j.chemosphere.2020.126396, https://linkinghub.elsevier.com/retrieve/pii/S0045653520305890

Stieger G, Scheringer M, Ng CA, Hungerbühler K (2014) Assessing the persistence, bioaccumulation potential and toxicity of brominated flame retardants: Data availability and quality for 36 alternative brominated flame retardants. Chemosphere 116:118–123, https://doi.org/10.1016/j.chemosphere.2014.01.083, http://dx.doi.org/10.1016/j.chemosphere.2014.01.083

Stroebe M, Scheringer M, Hungerbühler K (2004) Measures of Overall Persistence and the Temporal Remote State. Environmental Science & Technology 38(21):5665–5673, https://doi.org/10.1021/es035443s, https://pubs.acs.org/doi/10.1021/es035443s

Trudel D, Scheringer M, von Goetz N, Hungerbühler K (2011) Total Consumer Exposure to Polybrominated Diphenyl Ethers in North America and Europe. Environmental Science & Technology 45(6):2391–2397, https://doi.org/10.1021/es1035046, https://pubs.acs.org/doi/10.1021/es1035046

UN (2019) Globally Harmonized System of Classification and Labelling of Chemicals (GHS), eighth edn. https://www.unece.org/trans/danger/publi/ghs/ghs_rev08/08files_e.html

US EPA (1997) Exposure Factors Handbook. Tech. rep., https://cfpub.epa.gov/si/si_public_record_report.cfm?Lab=NCEA&dirEntryId=12464

US EPA (2019) EPA Releases First Major Update to Chemicals List in 40 Years. https://www.epa.gov/newsreleases/epa-releases-first-major-update-chemicals-list-40-years

Vandenberg LN, Colborn T, Hayes TB, Heindel JJ, Jacobs DR, Lee DH, Shioda T, Soto AM, vom Saal FS, Welshons WV, Zoeller RT, Myers JP (2012) Hormones and Endocrine-Disrupting Chemicals: Low-Dose Effects and Nonmonotonic Dose Responses. Endocrine Reviews 33(3):378–455, https://doi.org/10.1210/er.2011-1050, https://academic.oup.com/edrv/article-lookup/doi/10.1210/er.2011-1050

Wang Z, Boucher JM, Scheringer M, Cousins IT, Hungerbühler K (2017) Toward a Comprehensive Global Emission Inventory of C4-C10 Perfluoroalkanesulfonic Acids (PFSAs) and Related Precursors: Focus on the Life Cycle of C8-Based Products and Ongoing Industrial Transition. Environmental Science & Technology 51(8):4482–4493, https://doi.org/10.1021/acs.est.6b06191, http://pubs.acs.org/doi/abs/10.1021/acs.est.6b06191

Wang Z, Walker GW, Muir DCG, Nagatani-Yoshida K (2020) Toward a Global Understanding of Chemical Pollution: A First Comprehensive Analysis of National and Regional Chemical Inventories. Environmental Science & Technology https://doi.org/10.1021/acs.est.9b06379, https://pubs.acs.org/doi/abs/10.1021/acs.est.9b06379

Wegmann F, Cavin L, MacLeod M, Scheringer M, Hungerbühler K (2009) The OECD software tool for screening chemicals for persistence and long-range transport potential. Environmental Modelling & Software 24(2):228–237, https://doi.org/10.1016/j.envsoft.2008.06.014, https://linkinghub.elsevier.com/retrieve/pii/S1364815208001114

Risk Assessment and Management of Chemical Processes

7

7.1 The Aim of Process Risk Assessment

Chemical manufacturing involves producing often complex target molecules that have to be built up over multiple steps. This can mean that the necessary reaction and processing steps are carried out across different reactors and in multiproduct (or even multipurpose) production plants, potentially at large volumes and operating both day and night. Ensuring the safety of these processes is essential to prevent accidents and protect the health of the plant's employees as well as of the surrounding community and environment.

In the context of *process risk assessment*, a *chemical process* is considered to involve the production of one or more chemical products. This includes (1) the chemical reaction where raw materials (reactants) are converted to the desired product as well as to by-products and waste; (2) physical separation and purification operations such as crystallization, filtration, and distillation used to obtain the desired product specification; and (3) handling, transfer, and storage of chemicals.

Risks can exist due to the use of hazardous substances and conditions in chemical processes, as well as their susceptibility to a wide range of potential deviations. Critical deviations from planned operating procedures can stem from the technical equipment and controls involved, from human error during operation, or from environmental factors. They can result in severe accidents that have direct impacts on human health, the environment, and the equipment being used. These events include mainly fires, explosions, and releases of toxic substances into the atmosphere.

Process risk assessment aims to avoid, control, and mitigate critical process risks, with a special focus placed on the guiding principle of inherent process safety. Risk assessment and management should take place (1) at all stages during the development of new processes or installations and (2) for existing processes each time a modification to the process is made and as part of process safety management reviews. All identified process risks should be assessed and reduced to an acceptable level following societal values, protection goals, and considering costs.

Along the process development pathway, the amount of information, time, and effort needed for risk assessment increases from the research and development (R&D) stage toward the design, scale-up, and implementation stages of full-scale processes. The depth of the risk assessment needs to be systematically expanded based on the level of knowledge and data available. Consequently, the focus of the risk assessment shifts from the materials and reactions involved toward the processes and apparatuses used and lastly to assessing the control of the overall production plant including human factors (see Fig. 7.1).

Fig. 7.1 Main aspects involved in process risk assessment at different stages of development for a new process. A new process can involve the construction of a new plant, or it can be embedded in an existing one

Close interdisciplinary cooperation of experienced professionals during risk assessment is fundamental to ensure that it incorporates a high level of knowledge and experience. With such a team approach, risks are considered from different and broad perspectives, and unsuitable safety measures are detected in good time. The risk assessment team typically consists of the project developer, the future plant operator, a project engineer, a safety expert, and possibly other specialists. For a more extensive risk assessment, a trained moderator can be a useful addition to help the team methodically and efficiently consider all risk-relevant scenarios.

This chapter introduces a general procedure for the risk assessment of chemical processes; provides insight into relevant concepts, methods, and techniques; and shows their application in the form of a small case study. Further, detailed information on topics introduced in this chapter can be found in the multiple publications and textbooks developed by the Center for Chemical Process Safety (CCPS) (AICHE, 2018).

7.1.1   Relevant Terminology

Process risk assessment as introduced in this chapter involves a set of specific terminology. This section introduces some of the terms most often used, and these and others are further defined within the CCPS' Process Safety Glossary (CCPS, 2020).

The assessment of chemical process risks involves a review of accident scenarios, which consist of (1) *process hazards*, defined as an unplanned event or sequence of events that lead to a *loss event*, and (2) the *consequences* of the loss event. The concept of an accident scenario is illustrated in Fig. 7.2.

Within a process, a hazard is an inherent chemical or physical characteristic that has the potential to cause damage to humans, the environment, or property. Process hazards are determined by material properties, such as being flammable or toxic, and by process conditions, such as high temperature or high pressure.

An *event* by itself is the occurrence or change of a particular set of circumstances. And the consequence is defined as the undesirable result of an event. An event becomes a loss event when it is irreversible and results in adverse consequences. Examples in the chemical industry include the release of hazardous substances, fires, and explosions. A loss event can have several *initiating causes* as well as several consequences.

An initiating cause is the first event that occurs within potentially a sequence of events leading to a loss. It can be, for example, a process failure, a human error, or an external event, such as extreme weather conditions. It results in the transition from a normal situation to an abnormal situation, namely, in a *process deviation*.

Safeguards refer to any engineered system or administrative control used to interrupt the sequence of events following an initiating event (known as *preventive* safeguards), or to mitigate the consequences (known as *mitigative* safeguards). Some examples of safeguards include relief valves, automated control, operator responses, alarms, vents, and sprinklers.

7.1.2   Chemical Process Safety

The aim of process risk assessment and management is to achieve *chemical process safety*, which is based on the guiding principle of inherent safety during the process design (see Chap. 4) and on functional safety management (CCPS, 2016). Inherent process safety and functional safety are achieved through applying risk assessment and management throughout all stages of process development.

7.1.2.1 Inherent Process Safety

Inherently safe process conditions are achieved when process hazards have been permanently eliminated or reduced. To do this, process hazards should be identified during the early stages of process development. As defined by CCPS (2008b), four strategies for inherently safer design include:

1. Minimization of the material and energy contents in a process or plant
2. Substitution of hazardous materials or processes with non- or less hazardous alternatives
3. Moderation of hazardous process conditions, such as reduction of process pressure or temperature
4. Simplification of the design facilities and elimination of unnecessary complexity

Throughout this chapter, it should be kept in mind that inherent process safety is the primary guideline for chemical process risk assessment.

7.1.2.2 Functional Process Safety

Any process risks that remain after the implementation of inherently safer design can be further reduced by protection layers, which consist of safeguards not intrinsic to the process design. This is referred to as *functional process safety*. The general relationship between safeguards and loss events in an accident scenario is illustrated in Fig. 7.2 and further defined in ISO (2009).

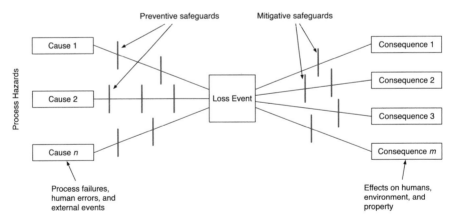

Fig. 7.2 Generic illustration of an accident scenario. Note that multiple causes leading to one or several consequences can occur simultaneously

The use of *protection layers* is an important tool within functional process safety and refers to the order (or sequence) in which safeguards are implemented in the system. The concept of protection layers is illustrated in Fig. 7.3 as defined in CCPS (2007), with inherently safe process design being at its core. While not all process risks can be eliminated during the design process, they can be reduced by implementing (multiple) layers of protection. These include controls (e.g., standard operating procedures, process alarms), supervision (e.g., protective alarms, operator monitoring), protections (e.g., safety instrumented systems), reductions (e.g., mechanical equipment such as pressure relief devices), barriers (e.g., explosion barriers), limiters (e.g., fire protection, emergency shutdown systems), and responses (e.g., evacuation to safe zones).

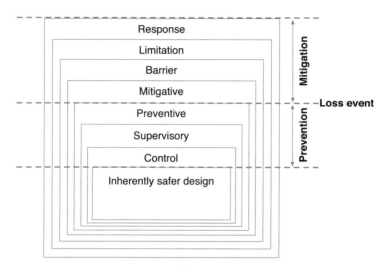

Fig. 7.3 Layers of protection for chemical processes as adapted from CCPS (2008a)

An *independent protection layer (IPL)* is a specific type of safeguard designed and managed to perform independently of any initiating cause or other layers of protection. Whether a protection layer is independent or not will have a significant influence on the risk assessment. IPLs have a higher-risk reduction potential than protection layers that are not independent.

A *safety instrumented system (SIS)* is a separate and independent combination of sensors, logic solvers, final elements, and support systems that are designed and managed to achieve a specified *safety integrity level (SIL)*. Integrity is a core attribute of a protection layer that relates to the achievable reduction of risk through the applied design and management of the layer.

An SIS could be, for example, designed to maintain safe operation of critical processes, such as activating cooling upon an unexpected increase in temperature. This function could be implemented by one or more protection layers.

Applying process risk assessment throughout all stages of process development helps to implement adequate protection layers across the entire system.

7.2   Procedure of Process Risk Assessment

Fig. 7.4 The six steps within chemical process risk assessment as adapted from ISO 31010 (ISO, 2009)

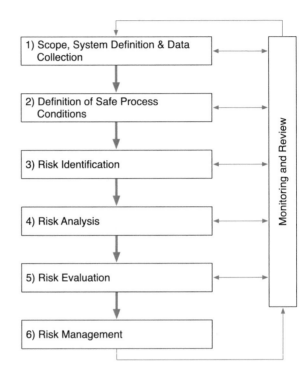

In the first step of a process risk assessment, the context of the risk assessment is established, and the required data are collected. The context consists of the scope of the assessment, the system definition, the available human and financial resources, and the selection of methods, techniques, and risk evaluation criteria. In the second step, safe process conditions for new processes are defined based on the physicochemical properties and chemical hazards of materials.

The first two steps provide the foundation needed to complete the actual risk assessment, which consists of risk identification (step 3), risk analysis (step 4), and risk evaluation (step 5). In the third step, potential accident scenarios along with their corresponding elements are systematically identified. During the risk analysis (step 4), the level of risk for the identified accident scenarios is characterized by analyzing their consequences and their probabilities of occurrence. In step 5, the acceptability of the risk is determined based on predefined criteria, and the final step of risk management takes action on any risks deemed not acceptable to eliminate or reduce them.

The results of a risk assessment performed at any given stage of process development are always transferred to the next development stage, and they are thus considered in every new risk assessment.

As process risk management is a proactive and preventive approach for ensuring process safety, it involves dealing with uncertainty. Correct modeling, recognition of relevant event scenarios, as well as reliability of the data used should always be critically reviewed (see also Crowl and Louvar (1990)).

7.3   Scope, System Definition, and Data Collection (Step 1)

7.3.1   Scope and System Definition

The *scope* of the process risk assessment and management needs to be defined considering societal, environmental, and economic aspects. This involves identifying and selecting the best methods and evaluation criteria to support the risk assessment, as well as assigning responsibilities and the necessary resources.

The scope also depends on the stage of process development. During the early stages of process development, the scope of assessment could be to compare different chemical synthesis routes and aspects of safety at different potential production sites, with the aim of promoting inherent process safety. At later stages, the aim could be instead to update a previous risk assessment to account for new knowledge. The selection of the exact analysis tools and evaluation criteria also depends on the defined scope of analysis and will be introduced in Sects. 7.5–7.7.

The *system boundaries* of the assessment also need to be defined and may limit the exercise to a specific location of a plant, piece of equipment, or single process. Decisions about whether to include connected utilities or adjacent processes also need to be taken. This might depend, for example, on the operation mode of a process being reviewed, namely, being *batch* or *continuous*. *Batch processes* follow a chronological execution of process steps according to a standard operating procedure and are produced in a specified amount within a set time frame. *Continuous processes* follow a location-dependent approach where the material flow is continuously in motion (without interruption) as it advances toward becoming the end product.

7.3.2   Data Collection

As mentioned previously, the required information for process risk assessment increases along the development pathway from research and development (R&D) toward the design, scale-up, and implementation of full-scale processes. Identifying which data should be collected at which point in time is a central task within integrated process development.

The data required for risk assessment of chemical processes can be classified as follows: (1) physicochemical properties and chemical hazard data for raw materials, intermediates, products, by-products, wastes, etc., (2) reaction and process conditions, (3) design specifications, (4) site data, and (5) management system data. Examples of these data types are presented below.

Physicochemical Properties and Chemical Hazard Data

- Composition, contaminants, etc.
- Physical state, volatility, solubility, etc.
- Thermal hazard data[1] such as the heat of reaction, reaction rate, generation of gases, etc. of the desired reaction, side reactions, consecutive reactions, and decomposition reactions.
- Chemical incompatibility or interaction matrices including the reactivity of process chemicals with one another, apparatus construction materials, utilities, the environment, contaminants, etc.
- Health hazards or human toxicity data such as LD_{50} values, maximum allowable workplace concentrations, carcinogenicity, etc. (see Sect. 6.3)
- Environmental hazard data, e.g., aquatic toxicity, bioaccumulation, biodegradability, etc. (see Sect. 6.3)
- Fire and explosion hazard data[2] such as flash point, explosion limits, ignition temperature, etc.

Some main sources of physicochemical properties and chemical hazard data include the scientific literature, chemical databases (such as SciFinder (Chemical Abstracts Service, 2018) and PubChem (Kim et al., 2016)), and published chemical encyclopedias (such as Ullmann's Encylopedia of Industrial Chemistry (Elvers, 2000)). Data obtained by experimental means are prioritized, whereas calculated methods such as quantitative structure-activity relationship models (QSARs) can help to fill remaining data gaps.

Reaction and Process Condition Data

- Process parameters such as temperature, pressure, concentration, pH, volume, reaction and decomposition time, etc.
- Phase transformations such as crystallization, melting, evaporation, sublimation, etc. and effects during these phase transformations such as delayed boiling, foaming, crust formation, etc.
- Electrostatic charge accumulation

Design Specification Data

- Process flow diagram with description of process steps, unit operations, and mass and energy balances
- Equipment specifications such as type, size, material, vacuum and pressure resistance, heating and cooling capacities, etc.

---

[1] The analysis of thermal hazards will be presented in Chap. 8.

[2] The definition of these properties and their relation to process safety are explained in Appendix C.

- Description of the control philosophy, process control system, safeguards, etc.

Site Data

- Supply systems for energy utilities and chemicals
- Waste treatment
- Fire protection systems
- Environmental influences at the site including heat, cold, rain, earthquakes, etc.

Management System Data

- Operating and maintenance procedures
- Internal standards and checklists
- Emergency response plan
- Results of previous risk assessment studies (if available)

7.4   Definition of Safe Process Conditions (Step 2)

This step enables the design of the process with the most appropriate conditions to achieve inherent process safety and reduce or eliminate process hazards. It defines *safe process conditions* that should be based on the physicochemical properties and chemical hazard data collected in the previous step, and these will form the basis for the subsequent steps of the risk assessment.

Any safe process conditions defined in a previous risk assessment(s) (e.g., applicable to processes at later stages of development) should be critically reviewed. This is important if new information is available, or if changes in the process have been implemented.

Some examples of how collected data can be applied to define safe process conditions include the use of:

- Health and environmental hazard data to establish critical threshold values for health and environmental protection
- Fire and explosion hazard data to identify the need for inerting, creation of an explosion protection zone, etc.[3]
- Chemical interaction matrices to indicate which substances should never come into contact with each other, the material that should be used for the processing equipment, and the substances that could react with utilities, air, and water
- Thermal hazard data for the design of the heat and cooling systems, charge volumes, etc. The analysis of thermal hazards will be introduced in detail in Chap. 8.

---

[3] How to establish safe process conditions in the presence of fire and explosion hazards is discussed in more detail in Appendix C.

Below, this step of the process risk assessment is illustrated using an example of establishing safe process conditions concerning fire and explosion hazards for the use of toluene.

7.4.1   Example: Defining the Safe Use of Toluene

Toluene is planned to be used in a set of manufacturing processes as a solvent. Using identified fire and explosion hazard data, safe process conditions for the use of toluene will be defined.

- Flash point: 6 °C
- Lower/upper explosive limit: 1.2/7.1 volume %
- Ignition temperature: 535 °C
- Minimum ignition energy: 0.26 MJ (at 25 °C), easily charged electrostatically
- GHS hazard statement H225: Highly flammable liquid and vapor
- GHS hazard statement H304: May be fatal if swallowed and enters airways

The low flash point indicates that flammable toluene/air mixtures can be formed at room temperature. Since toluene can be easily electrostatically charged and has low minimum ignition energy, electrostatic discharges can cause ignition. Since high electrostatic charge occurs especially in flowing media, careful attention should be paid to the pouring and transferring of toluene.

These precautions are also reinforced by the relevant GHS precautionary statements for handling toluene, which include:

- P210: Keep away from heat, hot surfaces, sparks, open flames, and other ignition sources. No smoking.
- P233: Keep container tightly closed.
- P240: Ground/bond container and receiving equipment.
- P243: Take precautionary measures against static discharge.

Considering this collected information, the safe use of toluene as a solvent in a manufacturing process should include setting the following process conditions:

- Use of a closed and inerted apparatus
- Setting up an explosion protection zone within the manufacturing location
- Earthing the equipment and using conductive tools and clothing

7.5   Risk Identification (Step 3)

This step is arguably the most important of the entire risk assessment since it identifies the risks that will be assessed. Risks not identified here will miss out on having measures set to avoid or mitigate them. The key question to answer during *risk identification* is: Where and how can systemic conditions become unsafe?

According to the risk management standards set by the ISO, risk identification is defined as the process of finding, recognizing, and recording risks. This identification involves recognizing potential *accident scenarios*, with all of their corresponding elements. As shown in Fig. 7.2, scenario elements include process hazards, the sequence leading up to a loss event, and the resulting consequences. If safeguards are present, they should also be considered. The output of this risk identification step can be thought of as a list of *m* potential consequences.

7.5.1 Methods for Risk Identification

Using on-site knowledge and experience is indispensable for the effective identification of risks. To systematically identify them, a wide range of methods are available and used within the chemical process industry. They each differ slightly in their assumptions, focus, complexity, and the results they deliver. While some of the methods are comprehensive and suitable for the identification of accident scenarios, others can be only used for the identification of process hazards or the detection of single equipment failures.

Two simple methods include *preliminary hazard analysis* and *checklist analysis*. Preliminary hazard analysis generates a list of process hazards based on hazardous properties (e.g., flammable or toxic material) and hazardous conditions (e.g., high-pressure reaction). Checklist analysis uses previously developed lists to verify the compliance of a system with standard practices, and identification of noncompliance can indicate the presence of process hazards.

Two other comprehensive and well-known methods used to predict accident scenarios are *what-if analysis* and *hazard and operability (HAZOP)* studies. What-if analysis is a brainstorming method in which experienced personnel focus on deviations from safe process conditions, designs, constructions, etc. that can lead to loss events. This method is suitable for nearly all stages of process development, with the required time and effort being proportional to the complexity and size of the system to be analyzed.

HAZOP studies also focus on deviations, but they use a more structured brainstorming approach. Here, a set of *guide words* are combined with specific process parameters and applied to specific operating steps to identify possible process deviations. Examples of guide words could be "no," "high," etc. Combined with the process parameters "flow" and "pressure," this can result in "no flow" and "high pressure." Tables 7.1 and 7.2 depict standard HAZOP guide words and some common HAZOP study process parameters. Table 7.3 shows some examples of deviations resulting from the combination of guide words and process parameters.

The HAZOP method requires considerable process knowledge as well as design and operation information, which makes it suitable for later stages of process development. The advantage of this method is its capacity to stimulate creativity and new ideas within the team, as well as its systematic approach. The use of both of these methods results in a list of scenarios, each with a unique combination of causes and consequences, considering the presence of safeguards.

Table 7.1 Original HAZOP study guide words, adapted from CCPS (2008a)

Guide words	Meaning
LESS or LESS OF	Quantitative decrease
MORE or MORE OF	Quantitative increase
NO or NOT	Negation of the design intent
PART OF	Qualitative decrease
AS WELL AS or MORE THAN	Qualitative increase
REVERSE	Opposite of the intent
OTHER THAN	Total substitution

Table 7.2 Common HAZOP study parameters, adapted from CCPS (2008a)

Speed	Level	Voltage	Temperature
Pressure	Composition	pH	Addition
Mixing	Viscosity	Frequency	Separation
Time	Flow	Information	Reaction

Table 7.3 Examples of deviations resulting from the combination of guide words and process parameters, adapted from CCPS (2008a)

Guide words		Parameter		Deviation
NO	+	MIXING	=	NO MIXING
MORE	+	TEMPERATURE	=	HIGH TEMPERATURE
AS WELL AS	+	ONE PHASE	=	TWO PHASE
OTHER THAN	+	OPERATION	=	MAINTENANCE

A third method that can be used is the *failure modes and effects analysis (FMEA)*, which focuses on individual equipment and system failures. This method analyzes how equipment can fail and how these failures can lead to process deviations and subsequently to loss events. FMEA considers each equipment failure as independent from other failures, except for the subsequent sequence of events resulting from a specific failure. Even though this method can be used to predict accident scenarios, it is not as comprehensive as the what-if analysis and the HAZOP study, since any initiating causes other than equipment failures will not be identified. The use of FMEA is especially suitable at later stages of process development for systems particularly vulnerable to single failures that can lead to loss events.

The selection of a risk identification method should be guided by the scope of the study. In some cases, it might be convenient to select one comprehensive method for the identification of accident scenarios, followed by the use of another method for identifying specific process failures.

7.5.2   Example: Application of Risk Identification Methods

To help illustrate how each of these risk identification methods can be used, they can be applied to the aqueous production process of ammonium acetate. In the process, an acetic acid solution (80%) and an ammonia solution (25%) are transferred continuously through flow valves to a water-cooled agitated reactor, as shown in Fig. 7.5. Ammonia and acetic acid react to form ammonium acetate. The aqueous ammonium acetate flows from the reactor to an open-top storage tank. Relief valves are provided on the storage tanks of ammonia and acetic acid as well as on the reactor, and they discharge to outside of the enclosed work area. The framework and process setup used here has been adapted from CCPS (2008a).

In this example, the focus is limited to considering what happens if too much ammonia is fed to the reactor (compared to the normal acetic acid feed rate). In this case, unreacted ammonia could carry over to the ammonium acetate storage tank. Any residual ammonia in the ammonium acetate tank will then be released into the work area, causing personnel exposure. Ammonia detectors and alarms are provided in the work area.

Tables 7.4, 7.5, 7.6, and 7.7 present examples of results obtained using the different methods introduced in this section. These results are not exhaustive and instead aim to provide an insight into how each method can be applied.

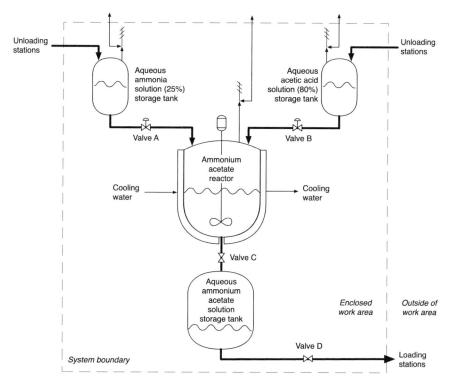

Fig. 7.5 Schematic of a production process of ammonium acetate from acetic acid and ammonia, adapted from CCPS (2008a)

Table 7.4 Example of using a checklist for risk identification of the ammonium acetate production process, adapted from CCPS (2008a)

Materials	Are all received materials individually inspected?	Yes, the identification of the truck itself and the paperwork for the shipment provided by the driver is confirmed before unloading takes place. However, no samples of the materials are taken, and the actual concentrations are not checked
Equipment	Are all equipment involved in the process inspected according to the maintenance schedule?	Yes, however a review of data on previous failures suggests that inspections specifically of the acid handling equipment may need to be done more frequently
Procedures	Do the process operators follow the set operating procedures?	No. New adjustments to operating steps have been recently implemented, but operators have reported that they believe one change could better protect the operator during the process step

Table 7.5 Example of applying the what-if analysis for risk identification within the ammonium acetate production process. A loss event that originated from two process deviations is shown. Adapted from CCPS (2008a)

Deviation	Process hazard	Loss event	Safeguards	Consequences
What if the concentration of acetic acid is too low?	Toxicity of ammonia	Unreacted ammonia could enter the ammonium acetate storage tank, be released into the work area, and expose operators[a]	Use of ammonia detector and alarm	Human health: skin corrosion, eye damage, death
What if valve B in Fig. 7.5 is closed?	Toxicity of ammonia	Unreacted ammonia could enter the ammonium acetate storage tank, be released into the work area, and expose operators	Routine maintenance; use of ammonia detector and alarm as well as a flow indicator in the acetic acid line	Human health: skin corrosion, eye damage, death

[a]Ammonia is a flammable gas that could explode if heated. This scenario is not considered in this example

7.6   Risk Analysis (Step 4)

In the *risk analysis* step, the potential consequences and the probability of occurrence of the previously identified accident scenarios are determined. As introduced in Chap. 2, risk, r, can be described as the product of the consequence, C, and the probability, P, of a particular scenario, s:

$$r(s) = C(s) \times P(s)$$

Table 7.6 Application of a HAZOP study for risk identification of the ammonium acetate production process. Here, one loss event that originated from two process deviations is shown. Note that these results are not exhaustive and that consequences on the environment are not considered. Adapted from CCPS (2008a)

Item	Process conditions	Deviation	Process hazard	Process failure	Loss event	Safeguards	Consequences
Acetic acid solution storage tank	Maintained at ambient temperature and atmospheric pressure, contains an 80% acetic acid feed solution with a tank level between 10% and 85% full	Low concentration of acetic acid (< 80%)	Toxicity of ammonia	(1) Low acetic acid concentration supplied by the vendor	Unreacted ammonia enters the ammonium acetate storage tank and is released into the work area	Ensure vendor is reliable; implement an acid unloading and transfer procedure and an ammonia detector with alarm	Human health: exposure can result in skin corrosion, eye damage, death
				(2) Error in loading acetic acid into the storage tank			
Acetic acid feed line to the reactor	Deliver 80% acetic acid feed solution to the reactor at the set flow rate and pressure	Low or no flow rate	Toxicity of ammonia	(1) No feed material in the acetic acid storage tank	Unreacted ammonia enters the ammonium acetate storage tank and is released into the work area	Ensure periodic maintenance; install ammonia detector with alarm and flow indicator in acetic acid line	Human health: skin corrosion, eye damage, death
				(2) Flow indicator fails			
				(3) Operator sets acid flow rate too low			
				(4) Acetic acid feed line control valve B fails			
				(5) Clogging of line			
				(6) Leak or rupture of line			

Table 7.7 Application example of an FMEA for risk identification of the ammonium acetate production process. Here, two loss events that originated from two failure modes of one equipment part are shown. These results are not exhaustive. Adapted from CCPS (2008a)

Equipment part	Failure mode	Failure effects (deviations)	Loss event	Safeguards	Consequences
Valve B regulating the acetic acid solution line	Fails and becomes closed	No flow of acetic acid to the reactor	Unreacted ammonia enters the ammonium acetate storage tank and is released into the work area	Ensure scheduled maintenance; implement an ammonia detector with alarm and a flow indicator in acetic acid line	Human health: skin corrosion, eye damage, death
Valve B regulating the acetic acid solution line	Valve ruptures	Significant amounts of acetic acid released into the enclosed work area	Workers exposed to high concentrations of acetic acid	Ensure scheduled maintenance; install valve designed for use with acids	Human health: skin corrosion, eye damage

Risk analysis can, therefore, be divided into the two parts of *consequence analysis* and *probability analysis*, and in both cases qualitative as well as quantitative approaches can be used. A qualitative approach within risk analysis relies on expert judgment, where the results are based on collective knowledge and experience of the risk assessment team. In a quantitative approach, this expert knowledge is considered in addition to quantitative approaches that use historical data as well as risk calculating techniques and models. This can help to reduce the subjectivity in the risk estimation; however, it still cannot avoid the uncertainty associated with the data used for calculations. In practice, the approach chosen is based on the scope of analysis and the availability of data and personnel resources.

7.6.1 Consequence Analysis

Consequence analysis assesses the severity of potential impacts associated with hazards inherent to the process design. Within this approach, the worst-case consequence of a loss event is considered, without taking into account the use of any mitigative safeguards.

Consequences from loss events can be classified into effects on (1) humans, (2) the environment, and (3) property. Often, they result primarily from the dispersion of toxic substances in the environment or from fires and explosions initiated by the release of hazardous material and energy via the process or storage equipment.

The focus of consequence analysis within process risk assessment is placed largely on instantaneous releases of gases and liquids, where release into the environment occurs within a relatively short time frame, e.g., in the case of the total rupturing of a reactor vessel (CCPS, 2009). In this subsection, the qualitative and quantitative approaches to consequence analysis are presented.

7.6.1.1 Qualitative Consequence Analysis

A qualitative consequence analysis can take the form of simple descriptions of effects on humans, environment, and property. The magnitude of each consequence can be described through categories using general terms such as "high," "medium," and "low." Any terminology used in the descriptions should be clearly explained. Table 7.8 shows an example of categories that can be defined for a qualitative consequence analysis with descriptive effects on people, the environment, and property.

Table 7.8 Example of consequence categories and descriptions for a qualitative consequence analysis. Adapted from ESCIS (1998)

Category	Description (effects on)		
	People	Environment	Property
Negligible	No effect	No effect	No effect
Low	Minor injury	Only equipment area involved	Minor damage to machinery
Medium	Injuries without irreversible damages	Working area involved, reversible damage in the neighborhood	Prolonged interruption of operations
High	Injuries with irreversible damages	Long-term damage also outside of the working area	Loss of a plant, loss of a building

7.6.1.2 Quantitative Consequence Analysis

In a quantitative consequence analysis, the effects on people, the environment, and property are quantified using models. Initially, *source term models* are often used to determine the rate and the total quantity (or duration) of the released materials (gases or liquids) and their thermodynamic states and consider the area in which the release takes place.

This information is subsequently used as input for dispersion, fire, and explosion models, which then are input into effect models used to estimate the resulting impacts on people, the environment, and property.

Source Term Modeling

The release of material or energy from a loss event depends on the physicochemical properties of the material, the process or storage conditions (e.g., compressed gas, pressurized liquefied gas, liquid), the mode of release (e.g., instantaneous or time-varying), and the interaction with the environment (e.g., aquatic toxicity) (CPD, 1996). Different types of models can be applied to estimate source terms that describe the release.

In the case of instantaneous releases of pressurized liquid gases, such as chlorine or ammonia, source term modeling includes modeling the evolution of the cloud size, the concentration in the cloud after the release (before being dispersed into the atmosphere), and the ratio of the material that remains in the air to the material that forms the evaporating liquid pool (see p. 2.114 in CPD (1996)).

For instantaneous non-boiling liquid discharges, pool spreading models are used to estimate source terms that are then incorporated into dispersion models or fire pool models (see p. 3.26 in CPD (1996)). For an instantaneous release of compressed gas, information about the release rate can be directly input into vapor cloud dispersion models.

Dispersion Modeling

Once the source terms of the released substance have been defined, dispersion models can be used to estimate the concentration in the atmosphere as a function of time and distance from the source.

The relevant independent variables for this calculation include the mean wind velocity, the released material volume (for instantaneous releases) or emission rate (for time-varying releases), the dimensions of the source, and the densities of the substance and the ambient air.

Particularly relevant for process risk assessment is the modeling of heavy gas dispersion, which describes the movement of released gases with a density greater than the ambient air. Many hazardous and volatile substances are heavier than air, including, for example, chlorine, ammonia, and hydrogen fluoride, which are released as cold gases (see p.4.42 in CPD (1996)).

A dense gas released into the atmosphere will descend to the ground and from there spread radially under the effect of gravity in a self-induced flow, producing a shallow cloud. For instantaneous releases of dense gases, wind dispersion has an important effect on the dilution of the gas cloud.

The CCPS distinguishes the following types of dispersion models: (1) phenomenological models or empirical relations; (2) intermediate models, such as box models; and (3) advanced models based on the Navier-Stokes equations. Detailed information about these models can be found in Appendix D as well as in Chap. 4 of CPD (1996).

Before running a detailed model, a preliminary screening can be carried out to estimate the concentration of a released substance near to the ground using the following formula:

$$c = \frac{Q}{u_{wc} \times H_{wc} \times W_{wc}} \tag{7.1}$$

- c: concentration [kg/m^3]
- Q: source emission rate [kg/s]
- u_{wc}: worst-case wind velocity [m/s] (assumed to equal 1 m/s)
- H_{wc}: worst-case cloud height [m] (assumed to equal 50 m)

- W_{wc}: worst-case cloud width [m] (assumed to equal $0.1\,x$, where x is distance from the source in meters)

The source emission rate (Q) is estimated by assuming that the total mass in the system is released over a 10-min period (i.e., 600 s) (see p. 7 in CCPS (1996)). If the estimated air concentration for the worst-case scenario is considerably lower than the set safe exposure level, then more detailed modeling may not be needed.

Fires and Explosions

The impacts of fires and explosions include overpressure and projectile effects, as well as thermal radiation effects, respectively. To quantitatively model these, vapor cloud explosion models and boiling liquid expanding vapor explosion models are used in the chemical industry.

A vapor cloud is formed when a large amount of flammable vaporizing liquid or gas is rapidly released into the atmosphere. This can result in a vapor cloud explosion or a flash fire if the cloud is ignited before being diluted below its lower flammability limit. After its release, if the vapor cloud comes into contact with an ignition source immediately after the cloud's release, then a flash fire will occur as the size of the cloud is sufficiently small. Otherwise, without an immediate ignition source, the accumulation of material in the cloud could increase to a concentration sufficient to cause an explosion (Santamaría Ramiro and Braña Aísa, 1998).

A boiling liquid expanding vapor explosion occurs when a large mass of pressurized superheated liquid (i.e., a liquid above its boiling point) or liquefied gas is released into the atmosphere. Such explosions are usually caused by a fire next to a storage container that weakens it and leads to a sudden shell rupture.

Further background information about fires and explosions is provided in Appendix C, and detailed information about these and additional approaches for modeling fires and explosions is available in CCPS (1999).

Effects on People, the Environment, and Property

In a final step of quantitative consequence analysis, outputs from the previous models are used to determine endpoint effects (damages) to people, the environment, and property. While the effects on people can be expressed in terms of fatalities and the effects on property in terms of monetary losses, the effects on the environment are more complex to assess and to express. Environmental effects from a loss event may involve contamination in multiple environmental compartments and include impacts in the surface water compartment on aquatic organisms such as algae, daphnia, fish, etc.

Consequential effects on people from exposure to toxic substances and/or fires and explosions can be estimated using a dose-response method. This can be coupled with a probit equation to linearize the response. Applying this *probit method* helps to describe a time-dependent relationship between a variable and its outcome defined by a normal distribution, and it results in a statistical correlation between a damage load or toxic dose and the percentage of affected people or property lost.

More information regarding effect modeling on people can be found in CCPS (1999). Section 6.3.2 also provides additional information on effect modeling for humans and the environment.

7.6.2　Probability Analysis

With the consequences of an accident scenario estimated, the second piece of information needed to define the risk is the likelihood of the accident occurring. This is estimated through *probability analysis*. In this step, the likelihood of an accident scenario is determined using the frequency of the initiating event and the probability of any preventive safeguards present within the process system failing.

As already mentioned, an initiating event can be the result of one or a combination of several process failures, human errors, or external influences. At very early stages of process development where no preventive safeguards are present, the *scenario likelihood* is equal to the frequency of the initiating event.

The probability of an accident can be reduced through the use of multiple safeguards. However, this will depend on whether there could be an independent failure of each safeguard, or if a single event could cause the failure of multiple safeguards simultaneously (known as a *dependent failure event*). An example of such a dependency could be the reliance on common support systems such as utilities, e.g., different equipment relying simultaneously on the same supply of electricity. If this single energy source fails, then multiple processes and their safeguards could all be compromised. If the potential for such dependent failure events is overlooked, the risk reduction measures implemented will be significantly overestimated. It is therefore extremely important to consider dependent failure events during probability analysis, and more detail specifically on this is provided in Appendix E.3.

The following sections present some general approaches for qualitative and quantitative probability analysis.

7.6.2.1 Qualitative Probability Analysis

Similarly to consequence analysis, the scenario likelihood can be qualitatively described within probability analysis using qualitative categories. An example of qualitative probability categories is shown in Table 7.9.

Table 7.9 Example of probability categories for a qualitative probability analysis

Category	Description
Negligible	Not anticipated to occur over the lifetime of the process or facility
Low	Anticipated to occur once over the lifetime of the process or facility
Medium	Anticipated to occur several times over the lifetime of the process or facility
High	Anticipated to occur more than one time each year

An example of a qualitative probability analysis for different initiating events (technical process failures, human errors, and external influences) according to expert knowledge is given in Table 7.10.

Table 7.10 Example of a qualitative probability analysis for different initiating events. Adapted from ESCIS (1998)

Probability	Technical failure	Human error	External influence
High	Analytical equipment (pH, redox, or O_2 probes)	Mix-up of products in similar packaging	Frost, rain
		Misinterpretation of verbal instructions	
Medium	Online measurement of data (pressure (p), temperature (T), level (L) sensors)	Mix-up of products delivered in drums/bags	prolonged power cut, transport accident
	Control valves	Misinterpretation of written working instructions	
Low	Independent elements	Confusion of products supplied through pipelines	Airplane crashes onto production facility
		Misinterpretation of written working instructions subjected to double checking	

7.6.2.2 Quantitative Probability Analysis

Several approaches can be applied to quantitatively assess the probability of an event. Following the definition of scenario likelihood presented earlier, this can be depicted using the expression:

$$F_{\text{scenario}} = F_{\text{initiating event}} \times P_{\text{safeguard failures}} \tag{7.2}$$

where F is the frequency of occurrence, usually given as the number of events per year, and P is the dimensionless probability. The scenario likelihood can in this way be expressed as the number of loss events per year.

Figure 7.6 illustrates that the probability can be quantitatively determined through either (1) directly using historical records if enough relevant data are available to provide statistically significant results, or, as is more often the case, (2) through using models such as *fault tree analysis* and *event tree analysis*. These models help to combine the probabilities of single events within the scenario sequence (e.g., from initiating events and safeguard failures) in a logical way.

In cases where they are not known, the probabilities of single events can also be modeled by examining equipment and human reliability, as well as using common cause failure analysis. More information about these techniques as well as about the fundamentals of probability can be found in Appendix E.

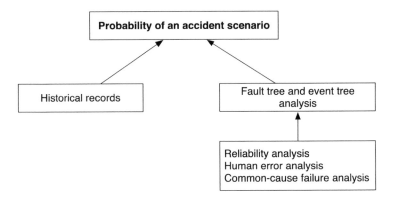

Fig. 7.6 Illustration of the different approaches that can be used for quantitative probability analysis of an accident scenario depending on the availability of data

Historical Data

If historical records of an accident scenario are sufficient, accurate, and applicable, they can be used to estimate its frequency of occurrence. Even though these data quality criteria are often difficult to fulfill, this technique can be applicable at the early stages of process development when other techniques cannot be applied (see p. 298 in CCPS (1999)). In this case, the frequency can be calculated by dividing the number of previous accidents caused by the total number of equipment in operation over a certain period. Using the example of a reaction vessel, if there have been 10 releases of a certain hazardous substance over a total of 1000 vessel-years,[4] the estimated release frequency will be 0.01 per vessel-year.

Fault Tree Analysis

Fault tree analysis (FTA) is a failure sequence model that can be used to estimate the probability of a loss event, known as the *top event*, that results from a sequence of earlier events, known as *causal events.*

A fault tree analysis starts from the top event and then identifies the causal events deductively. Each of the causal events (k) has an associated probability of occurrence P_k, which are usually estimated using reliability parameters. A tree diagram is used to graphically and logically order these events, and they are linked to one another via logical *AND* or *OR* operations.

---

[4]A vessel-year represents 1 year of a vessel being in operation. If there are 10 vessels which each have operated for 1 year, this is equal to 10 vessel-years.

AND operations within the tree signify that multiple initiating causal events (k) must occur simultaneously to trigger the event at the next higher level. This operator therefore reduces the total probability and can be defined as:

$$P = \prod_k P_k \tag{7.3}$$

where P is the total probability and P_k is the probability of event k.

OR operations within the tree mean that the occurrence of one of the associated causal events (k) is sufficient to trigger an event at the next higher level. The total probability can then be simplified as a sum of the individual event probabilities:

$$P = 1 - \prod_k (1 - P_k) \cong \sum_k P_k \text{ if } P << 1 \tag{7.4}$$

where P is the total probability and P_k is the probability of event k.

FTA makes the following assumptions: (1) all events are independent, (2) system components either perform successfully or fail completely, and (3) failures are instantaneous (i.e., time delays are omitted). Before using this technique it should always be checked whether these conditions are met. In cases where these conditions are not met, FTA can be adapted to include common cause failures.

Event Tree Analysis

Event Tree Analysis (ETA) is a method used to model the propagation of an initiating event that leads to many possible outcomes. Set up in a fashion that mirrors the fault tree analysis, in ETA the *top event* is considered the initiating event. Starting with the top event, each event following it is conditional on the occurrence of its precursor event. This form is also represented graphically by a tree.

The probability of occurrence of an outcome N (P_N) resulting from the initiating top event can be calculated as the product of the probability of the top event (P_{Top}) and all of subsequent conditional events along the event tree at each step k that leads to outcome N (P_{Nk}):

$$P_N = P_{Top} \times \prod_k P_{Nk} \tag{7.5}$$

In this way, an event tree can be used to estimate (1) the probability of occurrence of potential loss events that might result from a single initiating event, for example, after the loss of coolant, and (2) the probability of occurrence of consequences arising from a single loss event, for example, after the release of a hazardous substance.

Like FTA, event tree analysis primarily uses reliability data to define event probabilities. It also assumes that all events are independent, except for the preceding outcome branch, and that the system components either perform successfully or fail completely.

Combining Fault Tree and Event Tree Analyses

As shown in Fig. 7.7, *fault trees* and *event trees* can be combined to obtain a full picture of the causes and consequences of an accident scenario (as depicted earlier in Fig. 7.2).

In this case, the top event represents the loss event; the causal events of the fault tree represent process failures, human errors, or external influences; and the outcomes of the event tree represent the consequences of the loss event, namely, effects on people, the environment, and property. In this pathway of causes to consequences, preventive and mitigative safeguards are also considered.

Another related technique that can be used is the *bow tie analysis* (ISO, 2009). This approach provides a clear representation of accident scenarios and focuses especially on preventive and mitigative safeguards present in the system (see the structure of Fig. 7.2). However, unlike fault and event trees, bow tie analyses cannot be applied to cases where multiple causes occur simultaneously.

7.7 Risk Evaluation (Step 5)

Once the consequences and the probability of potential accident scenarios have been estimated, the *risk evaluation* step then combines these to determine the level of risk posed and classify it as acceptable or not acceptable. Acceptability can be determined through a comparison of the risk with protection goals set by internal corporate guidelines and by legislation, benchmarking with other processes, or through other predefined criteria. Risks deemed not to be acceptable will require subsequent safety improvement actions that will be identified and implemented in the next (and final) step of *risk management*.

One common way to estimate and communicate the level of risk is through the use of a risk matrix. These matrices help to visually estimate and present the level of risk using one axis to show increasing consequence severity and the other to show increasing probability of occurrence.

Consequences should include the estimated effects on people, the environment, and property, and its axis in the matrix can be represented through either a numerical scale or using descriptive terms such as "negligible," "low," etc. (see Table 7.8). Similarly, the probability axis can also be represented by a numerical scale or through descriptive terms (see Table 7.9).

Figure 7.8 provides a generic example of a risk matrix. In this example, the range of unacceptable risks is delimited by setting an *acceptability line*. Risks that exist above this line are classified as unacceptable and require risk management. The shading of cells within the matrix helps further visualize the level of risk and is set depending on how the probability and consequence scales are defined. The consequence and probability axes themselves can also be designed either symmetrically or to give more weight to one or the other.

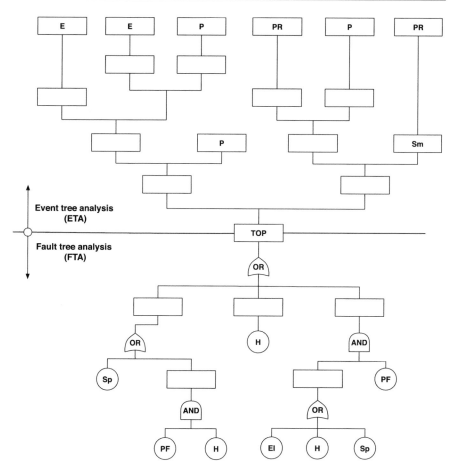

Fig. 7.7 Probability pathway from causes to consequences in an accident scenario by combining fault and event tree analysis. E = environment, P = people, PR = property, Sp = failure of preventative safeguard, Sm = failure of mitigative safeguard, PF = process failure, EI = external influence, H = human error

Other approaches can also be used to estimate and evaluate risks, such as risk indices for specific effects or plotting F-N curves.[5] For any approach used, evaluation of risks should be based on criteria that are as consistent and transparent as possible.

---

[5] An F-N curve is a plot of cumulative frequency versus consequences (often expressed as the number of fatalities) (CCPS, 2020).

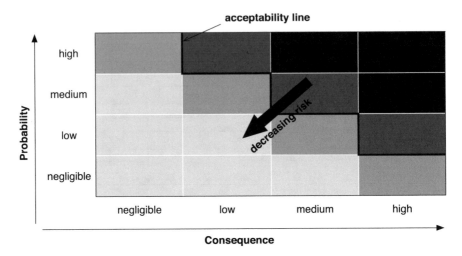

Fig. 7.8   Example of a qualitative risk matrix

7.8   Risk Management (Step 6)

In the final step of process risk assessment, *risk management* decisions are made on how to control any risks classified as unacceptable during the previous evaluation step. Safety measures need to be selected and implemented to reduce risks to an acceptable level, which usually involves the following actions (in order of decreasing priority):

• Elimination of potential hazards and selection of a process with the lowest risk according to the guiding principle of inherent safety
• Implementation of layers of protection (functional safety):
    – Safety Instrumented System (SIS)
    – Organizational and personnel measures
    – Preparation of an emergency response plan

When planning safety measures for risk management, the following general aspects should be considered:

• Safety measures should be carefully checked for their introduction of any new risks (e.g., when installing a safety valve to manage the pressure of a reactor, there is a new risk for the potential release of toxic or explosive substances).
• The optimal interaction between humans and machines is especially important. In a fully automated operation, operators are controlled by technology. However, in the case of a failure, the technology should be under the control of the operators.

- Safety measures should be designed so that a simple error cannot lead to an event with severe consequences.
- The benefits of a risk analysis are highly dependent on the careful implementation and subsequent preservation of the selected protective measures.
- The goal of safety measures also includes consideration of a cost-benefit ratio (see Sect. 7.8.1).

After the implementation of safety measures, the level of risk should be assessed once again to determine if additional safety measures are required.

7.8.1   Decision-Making During Risk Management

The decisions made during the risk management step are subjective and depend on factors such as the set protection goals, the corporate safety culture, existing legislation, technical safety norms, the current state of technology, and cost-benefit considerations of treating the risk.

A common approach for decision-making is the "as low as reasonably practicable" (ALARP) criteria system, which consists of dividing risks into three bands: (1) an upper band where the level of risk is considered unacceptable independently of the benefits associated with the activity, and risk treatment is required independently of the costs, (2) a middle band (or "gray" area) where the costs and benefits of increasing safety and of accepting risk are compared through a *cost-benefit ratio*, and (3) a lower band where the level of risk is considered to be low enough that no risk treatment measures are needed (ISO, 2009).

The concept of a cost-benefit ratio is illustrated in Fig. 7.9. It shows that reducing a risk through safety measures results in lowering the annual expected damage costs. However, it also results in an exponential increase in costs for safety measures. All costs are expressed in monetary amounts [$/year]. An economically optimal level is reached when the sum of avoided damage costs and the costs for safety measures reaches a minimum. The damage costs decrease linearly; however, the increase in costs for implementing safety measures is not linear. The marginal utility of implementing safety measures to reduce risks and avoid damage costs ($\frac{\Delta \text{Damage Costs}}{\Delta \text{Costs for Safety Measures}}$) therefore steadily decreases with increasing demand for safety. Eventually, one dollar invested in improving safety will result in significantly less than one dollar worth of reductions in damage costs.

7.8.2   Residual Risk

Even after the implementation of safety measures to reduce the risk, a certain amount of risk will always remain. This is known as *residual risk* and includes:

- risks that are accepted as a result of cost-benefit considerations

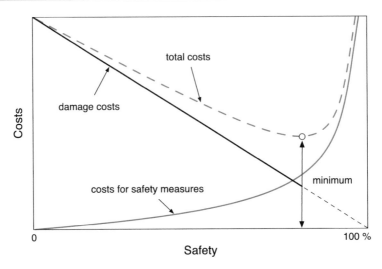

Fig. 7.9 Illustration of a cost-benefit ratio showing damage costs, investment costs in safety measures, and total costs attributed to managing a risk

- risks that are incorrectly assessed due to uncertainties arising from the data, modeling, and assumptions
- risks that exist but have not been identified (perhaps the most important to consider)

There is no general measure or threshold that can be used to determine how acceptable a residual risk is. Residual risks should be managed through a periodic review of the risk assessment results along with its corresponding assumptions, critical scenarios, and uncertainties. Every opportunity should be taken to reduce these and to further improve the quality and validity of the assessment. In addition to the purely technical and economic aspects, social aspects should also be carefully considered in evaluating residual risks, which could require facilitating a risk-benefit dialogue with the potentially affected population (see Chap. 9). Having clear and complete documentation of the risk assessment also helps to justify the safety measures implemented and the residual risks accepted during a future review. This can be especially important to have when responding to legal inquiries in the case of an accident.

As a whole, process risk assessment can be an intensive undertaking with a wide range of scenarios to carefully consider and address. In general, a process can be considered safely designed and managed if:

- a thorough risk assessment has been carried out
- the safety measures (risk management) taken:
    - correspond to the findings of the risk assessment
    - take into account the current state of knowledge of science and technology

– are compliant with the protection goals of the company and society
– have been rigorously implemented and are regularly inspected

7.9   Example: Wet Grinding of Cyanuric Chloride

In this section, the procedure of process risk assessment is illustrated through a simplified application to a process step within the production of a reactive dye. This particular process step involves the grinding of cyanuric chloride in an aqueous medium to bring it into a reactive form necessary for the subsequent reaction steps.

A known hazard of this process step is the hydrolytic decomposition of cyanuric chloride as shown in Fig. 7.10. Cyanuric chloride decomposes in water via hydrolysis to produce hydrochloric acid (HCl) and the final product cyanuric acid. Important to note is that this reaction is *autocatalytic*. Each of the three acid chlorides shown in Fig. 7.10 will be hydrolyzed, which results in the production of HCl, the lowering of the pH, and therefore an increasing reaction rate. This autocatalytic nature makes the reaction even more dangerous and is an important aspect of the risk assessment to consider.

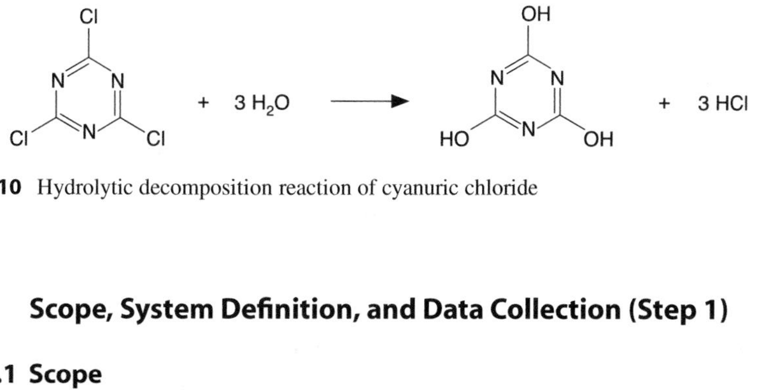

Fig. 7.10   Hydrolytic decomposition reaction of cyanuric chloride

7.9.1   Scope, System Definition, and Data Collection (Step 1)

7.9.1.1 Scope
The *scope* of this process risk assessment is to identify, evaluate, and reduce the risk of the hydrolytic decomposition reaction of cyanuric chloride during the wet grinding process at an early stage of process development.

7.9.1.2 System Definition
The system boundaries, in this case, are defined by the batch reactor where the grinding process takes place. Energy carriers (e.g., water, ice) and their corresponding pipelines are considered within the system boundaries, but the storage of cyanuric chloride is not. This system is illustrated in the schematic shown in Fig. 7.11.

Fig. 7.11 System considered within the risk assessment of the grinding process of cyanuric chloride (CC). Note that steam is not used in the process, but it is part of the standard equipment connected to the reactor and therefore considered as part of the system. RST = rotor/stator turbine as grinder

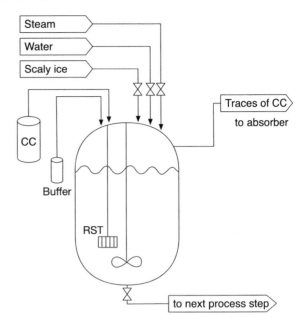

7.9.1.3 Data Collection

Reaction and Process Conditions

193 kg (1.05 kmol) of the poorly water-soluble cyanuric chloride is placed in suspension in an aqueous buffered medium with ice (comprising 250 kg water, 5 kg Na_2HPO_4 (25%), and 250 kg ice) and finely ground in a batch reactor[6] at a temperature of 0–2 °C and a pH of 7.

Under these process conditions (temperature and pH), the hydrolysis of cyanuric chloride proceeds very slowly and can be considered as noncritical for at least 12 h (Yan et al., 2008). Through an assessment of the thermal hazards, which will be introduced in more detail in Chap. 8, we know that at higher temperatures and a lower pH, the acid-catalyzed decomposition proceeds autocatalytically due to the HCl formed. A thermal runaway reaction could then occur after 10–20 min when the reaction mixture is heated to 50 °C and at a pH of 5–7.

Physicochemical Properties and Chemical Hazard Data

Upon review of property and safety data describing the chemicals used in the process step, the following information was found:

---

[6]Note that this is a mechanical process where no chemical reaction should take place.

Cyanuric Chloride

- Molecular weight: 184 g/mol
- Water solubility at 25°C: 44 mg/L
- Reaction enthalpy change from hydrolysis: -450 kJ/mol

Buffer (Na_2HPO_4)

- Health hazards: minimal to no effects found
- Environmental hazards: minimal to no effects found

Hydrochloric Acid (HCl)

- Molecular weight: 36.5 g/mol
- Health hazards: causes severe skin burns and eye damage, causes respiratory irritation above set maximum exposure level[7] of 0.03 g/m^3 air
- Environmental hazards: minimal to no effects found

Cyanuric Acid

- Health hazards: causes serious eye, skin, and respiratory irritation
- Environmental hazards: minimal to no effects found

Interaction Matrix of the Grinding Process

Table 7.11 shows an interaction matrix between chemicals involved in the process step. It depicts which combinations of the chemicals react with one another and which of those reactions are hazardous. The materials of the equipment in the process that come into contact with the chemicals are not taken into account in this matrix; however, an exhaustive assessment should consider them.

Table 7.11 Interaction matrix for process chemicals and the decomposition product HCl. Nonhazardous reactions that result are marked with "(+)." Hazardous reactions are marked with "+"

Chemical substance	CC	Water	Na_2HPO_4	HCl
CC				
Water	+		(+)	(+)
Na_2HPO_4		(+)		(+)
HCl		(+)	(+)	

---

[7]Maximum air concentration below which health effects are not expected to occur that limit a person's ability to seek protection (US EPA, 2009).

Design Specifications

The following design specifications for the process have currently been set:

- Reactor volume: 1 m^3
- Construction material: rubber-coated steel vessel
- Process makes use of an:
    - Anchor agitator
    - Polytron grinder
    - Drum for cyanuric chloride
    - Drum for Na$_2$HPO$_4$
    - Feed lines for steam, ice, and water (three separate pipelines)
    - pH and temperature measurement devices

The standard operating procedure for the process step defines the following sequence of events:

1. Check lines; close floor valve
2. Add water
3. Turn on stirrer
4. Add buffer
5. Add ice
6. Add cyanuric chloride
7. Turn on grinder
8. Stir and grind
9. Turn off grinder
10. Check lines; open floor valve and drain into the next reactor (condensation with amine)
11. Turn off stirrer
12. Wash reactor with water

7.9.2   Definition of Safe Process Conditions (Step 2)

Considering the data collected in the first step of the risk assessment, the following safe process conditions can be defined:

- Temperature of 0–2 °C
- pH of approximately 7
- Controlled and continuous stirring
- Adherence to the sequence of loading as defined in the standard operating procedure
- Absorption of any traces of cyanuric chloride
- Use of a rubber-coated steel vessel

7.9.3 Risk Identification (Step 3)

In this example, risk identification is carried out using a HAZOP study. As introduced in Sect. 7.5.1, this technique focuses on identifying deviations and uses guide words combined with specific process parameters.

Table 7.12 presents the results of the HAZOP analysis. The guide words used to address potential deviations in this process step are "high temperature," "low pH," and "no mixing." In this example, the loss event resulting from these deviations is identified as being the hydrolytic decomposition of cyanuric chloride. An autocatalytic runaway reaction, thermal explosion, and subsequent release of HCl in the environment are the worst-case consequences of these deviations. The hazards inherent to the process, as well as potential process failures (causes), and existing preventive and mitigative safeguards are also shown in Table 7.12. Note that this table does not present the results of a truly exhaustive HAZOP study.

7.9.4 Risk Analysis (Step 4)

Within the risk analysis step, the identified consequences from the HAZOP study are further investigated along with their probabilities of occurrence.

7.9.4.1 Consequence Analysis

From the HAZOP study, a worst-case consequence from a decomposition reaction of cyanuric chloride and the failure of safeguards would be a thermal runaway, explosion, and the resulting release of hydrochloric acid (HCl) into the atmosphere. To help analyze the consequences of the HCl release, Eq. 7.1 can be used to estimate the distance from the release source (x) that the vapor cloud would need to travel before its concentration (c) falls below the safe maximum exposure levels. The risk of a thermal runaway specifically for this example will be presented in Sect. 8.4 in Chap. 8.

According to stoichiometric calculations, it is known that upon complete hydrolysis of the defined batch size of cyanuric chloride (1.05 kmol), approximately 111 kg (or 75 m^3) of HCl is produced. Assuming that this amount is released within 10 min, that the wind velocity is equal to 1 m/s, that the cloud height is 50 m, and that the cloud width is $0.1 \times x$ (where x is the distance from the source) and using the set maximum and critical exposure level as the concentration (c), Eq. 7.1 can be set up as:

$$0.03 \frac{g}{m^3} = 3 \times 10^{-5} \frac{kg}{m^3} = \frac{111 \text{ kg}/600 \text{ s}}{1 \frac{m}{s} \times 50 \text{ m} \times 0.1 \times x \text{ m}}$$

which can be solved for x as:

$$x = 1233 \text{ m}$$

Table 7.12 Results of a HAZOP study of the wet grinding process of cyanuric chloride

Item	Process conditions	Deviation	Process hazard	Process failure	Loss event	Safeguards	Consequences
Batch reactor for grinding process of cyanuric chloride	pH of approximately 7	Lower pH	(1) Autocatalytic decomposition reaction of cyanuric chloride	(1) Failure in buffer supply	Decomposition of cyanuric chloride	(1) Buffer (prevention)	(1) Thermal runaway and thermal explosion
			(2) Toxicity hazards (cyanuric chloride, HCl)	(2) Failure of pH measurement		(2) pH alarm (prevention)	(2) Release of hydrochloric acid (HCl) from the process
				(3) Failure of electricity supply		(3) Depressurization (mitigation)	
						(4) Rupture disk activation (mitigation)	

Batch reactor for grinding process of cyanuric chloride	Temperature = 0–2 °C	Higher temperature	(1) Autocatalytic decomposition reaction of cyanuric chloride	(1) Failure in the ice supply	Decomposition of cyanuric chloride	(1) Ice (prevention)	(1) Thermal runaway and thermal explosion
			(2) Toxicity hazards (cyanuric chloride, HCl)	(2) Accidental injection of steam in the reactor (valve accidentally opened or damaged)		(2) Temperature alarm (prevention)	(2) Release of hydrochloric acid (HCl) from the process
				(3) Failure of temperature measurement		(3) Depressurization (mitigation)	
				(4) Failure of electricity supply		(4) Rupture disk activation (mitigation)	
Batch reactor for grinding process of cyanuric chloride	Stirring	No mixing (poor heat exchange)	Tendency of cyanuric chloride to agglomerate (local temperature and pH gradient)	Failure of electricity supply	Decomposition of cyanuric chloride	Stirrer alarm	(1) Thermal runaway and thermal explosion
							(2) Release of hydrochloric acid (HCl) from the process

This result shows that dangerous concentrations of HCl will travel more than 1 km from the emission source, which is outside of the working area and into the surrounding community. Considering this and the toxicity of HCl, the effect on people is determined as being "high." For the thermal explosion, the effects on people and property are also classified as "high."

The results of the qualitative consequence analysis are summarized in Table 7.13 using the consequence categories as defined in Table 7.8.

Table 7.13 Results of the qualitative consequence analysis for the grinding process of cyanuric chloride

Consequence	Effects on people	Effects on environment	Effects on property
Thermal runaway and thermal explosion	High	Low	High
HCl release	High	Low	Medium

7.9.4.2 Probability Analysis

As introduced in Sect. 7.6.2, the probability of occurrence is dependent on the probability of the initiating event(s) (process failures, human errors, external influences) and the probability of failure of any preventive safeguards present in the process system.

Table 7.14 shows the results of a qualitative probability analysis for this process step using the probability classes defined in Table 7.9 and the estimations based on expert knowledge provided in Table 7.10. Probability classes are first assigned to process failures that lead to process deviations and then to the failures of preventive safeguards (included in Table 7.12). Based on the contribution of these probabilities, the probability of the loss event (decomposition of cyanuric chloride) is estimated to be "medium."

The probability of a thermal runaway is estimated using thermal safety techniques and results in an estimated probability of "medium." These concepts will be introduced in Chap. 8, and thermal safety calculations specifically for this example will be presented in Sect. 8.4.

The mitigative safeguards are the depressurization valve and rupture disk as shown in Table 7.14. The failure probabilities of these safeguards are estimated to be "medium" and "low", respectively. Taking these probabilities into consideration as well as the probability of the loss event, the overall scenario probability is estimated to be "medium."

Here, the focus is placed on using a qualitative analysis; however, a quantitative probability analysis can also be carried out with the methods of *fault tree analysis* (FTA) and *event tree analysis* (ETA). The application of these methods to this example is shown in Appendix E.4.

Table 7.14 Qualitative probability analysis for the process step of grinding cyanuric chloride. The qualitative estimate of probability for each entry is shown in italics

Process failure	Process deviation	Safeguard failure (prevention)	Loss event	Safeguard failure (mitigation)	Consequence
Failure in the ice supply (*medium*)	Temperature too high (*medium*)	Failure of temperature alarm (*medium*)	Decomposition of cyanuric chloride (*medium*)	Temperature/ pH stabilization (*medium*)	Thermal runaway and thermal explosion (*medium*)
Accidental injection of steam in the reactor (*low*)	pH too low (*medium*)	Failure of pH alarm (*high*)		Ice quenching (*medium*)	HCl release (*medium*)
Failure of temperature measurement (*medium*)		Temp./pH stabilization (*high*)		Depressurization (*medium*)	
Failure in buffer supply (*medium*)				Rupture disk activation (*low*)	
Failure of pH measurement (*high*)					
Failure in the electricity supply (*medium*)					

7.9.5   Risk Evaluation (Step 5)

In the next step, the level of risk of the grinding process of cyanuric chloride can be evaluated by combining the results of the consequence and probability analyses obtained. This can be done by using a risk matrix.

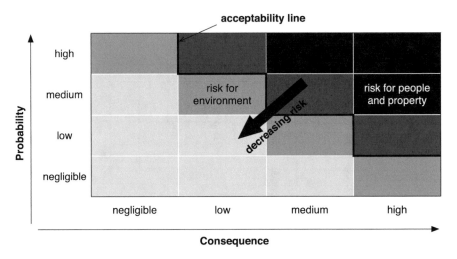

Fig. 7.12 Risk matrix for the wet grinding of cyanuric chloride

As shown in Fig. 7.12, the plotted level of risk concerning effects on people and property is above the set acceptability line. It is therefore classified as being unacceptable, and this signifies that risk reduction measures need to be implemented.

7.9.6   Risk Management (Step 6)

Given the need for risk reduction, appropriate measures have to be identified, selected, and implemented to bring the risks for people and property down to an acceptable level. Some of the risk reduction measures that could be implemented for this process step include:

- Installing a blind flange in the steam pipe or securing the manual valve with a safety lock. This eliminates the risk of accidental steam injection into the process and illustrates the principle of inherent process safety.
- Installing the option for pneumatic stirring. This reduces the risk in case of a failure in the electricity supply and is an example of a preventive safeguard.
- Installing an alkaline adsorption column in case of HCl release. This is an example of a mitigative safeguard.

These risk reduction measures should be considered also in terms of their feasibility to implement as well as their cost-benefit ratios. Once specific measures are selected, the level of risk needs to be reassessed to ensure the process risk would reach acceptable levels after their implementation. Even after the implementation of these and potentially other risk management measures, remaining residual risks should be considered. These could include, for example, a simultaneous failure of all electricity sources that deactivates measurement devices and alarms as well as a mechanical failure of the alkaline absorption column. These should be monitored and further reduced in future risk assessments that make use of new data, knowledge, and experience within the risk management team.

References

AICHE (2018) Center for Chemical Process Safety. URL https://www.aiche.org/ccps

CCPS (1996) Guidelines for Use of Vapor Cloud Dispersion Models, 2nd edn. URL https://www.aiche.org/ccps/publications/books/guidelines-use-vapor-cloud-dispersion-models-2nd-edition

CCPS (1999) Guidelines for Chemical Process Quantitative Risk Analysis, 2nd edn. John Wiley & Sons, Inc., URL https://www.wiley.com/en-us/Guidelines+for+Chemical+Process+Quantitative+Risk+Analysis%2C+2nd+Edition-p-9780816907205

CCPS (2007) Guidelines for Safe and Reliable Instrumented Protective Systems. John Wiley & Sons, Inc.

CCPS (2008a) Guidelines for Hazard Evaluation Procedures. John Wiley & Sons, Inc., Hoboken, NJ, USA, https://doi.org/10.1002/9780470924891, URL http://doi.wiley.com/10.1002/9780470924891

CCPS (2008b) Inherently Safer Chemical Processes. John Wiley & Sons, Inc., Hoboken, NJ, USA, https://doi.org/10.1002/9780470925195, URL http://doi.wiley.com/10.1002/9780470925195

CCPS (2009) Guidelines for Developing Quantitative Safety Risk Criteria. John Wiley & Sons, Inc., URL https://www.wiley.com/en-us/Guidelines+for+Developing+Quantitative+Safety+Risk+Criteria-p-9780470261408

CCPS (2016) Guidelines for Safe Automation of Chemical Processes, 2nd edn. John Wiley & Sons, Inc., Hoboken, NJ, USA, https://doi.org/10.1002/9781119352044, URL http://doi.wiley.com/10.1002/9781119352044

CCPS (2020) CCPS Process Safety Glossary. URL https://www.aiche.org/ccps/resources/glossary

Chemical Abstracts Service (2018) SciFinder. URL https://www.cas.org/products/scifinder

CPD (1996) Methods for the calculation of Physical Effects. Tech. rep., URL http://content.publicatiereeksgevaarlijkestoffen.nl/documents/PGS2/PGS2-1997-v0.1-physical-effects.pdf

Crowl D, Louvar J (1990) Chemical Process Safety: Fundamentals with Applications, 1st edn. Prentice-Hall, Englewood Cliffs

Elvers B (ed) (2000) Ullmann's Encyclopedia of Industrial Chemistry. Wiley, Weinheim, Germany, https://doi.org/10.1002/14356007, URL http://doi.wiley.com/10.1002/14356007

ESCIS (1998) Introduction to Risk Analysis. ESCIS Safety Series, 4. URL https://doi.org/10.3929/ETHZ-B-000354485

ISO (2009) ISO 31010 Risk management - Risk assessment techniques. Tech. rep., URL https://www.iso.org/standard/72140.html

Kim S, Thiessen PA, Bolton EE, Chen J, Fu G, Gindulyte A, Han L, He J, He S, Shoemaker BA, Wang J, Yu B, Zhang J, Bryant SH (2016) PubChem Substance and Compound databases. Nucleic Acids Research 44(D1):D1202–D1213, https://doi.org/10.1093/nar/gkv951, URL https://doi.org/10.1093/nar/gkv951

Santamaría Ramiro JM, Braña Aísa PA (1998) Risk Analysis and Reduction in the Chemical
    Process Industry. Springer Netherlands, Dordrecht, https://doi.org/10.1007/978-94-011-4936-
    5, URL http://link.springer.com/10.1007/978-94-011-4936-5
US EPA (2009) Risk Management Program Guidance for Offsite Consequence Analysis. Tech.
    rep., URL https://www.epa.gov/sites/production/files/2013-11/documents/oca-chps.pdf
Yan Z, Xue WL, Zeng ZX, Gu MR (2008) Kinetics of Cyanuric Chloride Hydrolysis in Aqueous
    Solution. Industrial & Engineering Chemistry Research 47(15):5318–5322, https://doi.org/10.
    1021/ie071289x, URL https://pubs.acs.org/doi/10.1021/ie071289x

Thermal Process Safety

8

8.1　Thermal Process Safety as a Part of Process Risk Assessment

To be economically efficient, chemical production on an industrial scale often requires high productivity through the use of large production volumes and high concentrations. Consequently, this can be associated with the generation of considerable amounts of thermal energy too large to be absorbed by the environment surrounding the system. The topic of *thermal process safety* aims to address specifically this key hazard that can exist within many chemical processes.

Instead of having its own set of specific steps, it is rather embedded within the more general framework of chemical process risk assessment introduced in Chap. 7. As thermal process safety is fundamental for the design of inherently safe processes and the definition of safe process conditions during the early stages of process development, this chapter is dedicated specifically to introducing this topic and some of the most important physicochemical phenomena that shape it.

In the field of thermal process safety, the basic principles of thermodynamics, chemical kinetics, and reaction engineering are used to identify thermal hazards and define safe process conditions. Here, a *thermal hazard* is referred to as the potential of a chemical system to cause a significant increase in temperature and/or pressure that cannot be absorbed by its surroundings. This can be the result of, for example, highly exothermic decomposition reactions.

One of the most fundamental concepts related to thermal hazards is the *thermal runaway*. A thermal runaway is an uncontrolled reaction that acts as a feedback loop. Strong exothermic reactions are accelerated by a temperature rise, which leads to yet further and more rapid increases in temperature and pressure. Left unchecked, these reactions can lead to a thermal explosion. Preventing such *runaway* reactions is therefore clearly of significant relevance to chemical process safety.

Runaway reactions are the result of the insufficient removal of heat, normally from large units such as reactors and storage vessels on an industrial scale. In storage

© The Author(s), under exclusive license to Springer Nature Switzerland AG 2021
K. Hungerbühler et al., *Chemical Products and Processes*,
https://doi.org/10.1007/978-3-030-62422-4_8

vessels, they can be initiated, for example, by external heat sources or through contamination within the tank itself. During a production process in a reactor, runaway reactions can arise from the heat produced by the desired reaction or by undesired decomposition reactions set off by the accumulation of reactants, lack of a cooling medium, loss of agitation, etc.

Applying thermal process safety within process risk assessment involves identifying and managing the *thermal risks* that are present, which includes a characterization of (1) the energy potential of the chemicals involved in the process, (2) their potential reaction and decomposition rates, and (3) the implications for the process equipment. This normally requires the review of substance data (e.g., physicochemical properties), the use of estimation methods (e.g., QSARs based on the molecular structure), thermodynamic and kinetic calculations (e.g., heat balances), and experimental measurements (e.g., testing of thermal stability).

In this chapter, Sect. 8.2 introduces some of the fundamental concepts of thermodynamics and kinetics related to reactor stability. Then, Sect. 8.3 (1) introduces important thermokinetic concepts, (2) presents the scenario of a thermal runaway as result of a cooling failure during an exothermic batch reaction,[1] and (3) analyzes thermal risks in terms of impact and probability and evaluates the level of risk. Finally, Sect. 8.4 applies these concepts within the process risk assessment of the grinding of cyanuric chloride, which was first introduced in Chap. 7. For further, in-depth explanations of these concepts and additional examples, see the textbook developed by Stoessel (2008).

8.2   Heat Balance and Reactor Stability

Ensuring safe process conditions for a chemical process is based on a good understanding of its *heat balance*. Some of the fundamental characteristics that define a chemical process and directly affect the generation and management of its heat include:

- the mode of the reaction (e.g., batch, semi-batch, continuous)
- the type of mechanism used for heat removal (e.g., external cooling via reactor wall, melt, or evaporative cooling, etc.)
- the choice of operating parameters (e.g., temperature, concentration, dosing time, stirring intensity, etc.)
- the design of the reactor (e.g., specific cooling surface, stirrer, reactor material, etc.)

---

[1] Most of the chemical reactions performed in the fine chemical industry are exothermic.

In this chapter, a main focus will be placed on introducing and understanding heat balances within *batch reactor* processes. Compared to other types of ideal reactors (e.g., semi-batch reactors, ideally stirred tank reactors, plug flow reactors, etc.), batch reactors can result in some of the most critical impacts. Nevertheless, most of the concepts illustrated here can be applied to other types of chemical reactors and storage systems.

8.2.1   The Batch Reactor

In a batch reaction process, all reactants are fed into the reactor vessel at once, and the start of the reaction is carried out by increasing the temperature, making use of a catalyst, or adding reaction components simultaneously. Figure 8.1 shows a batch reactor, where reactants A and B react to produce product P.

Fig. 8.1 Schematic of a batch reactor with reactants A and B that produce product P

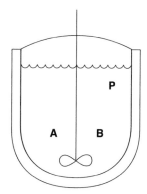

The simplified heat balance for a batch reactor, neglecting heat losses and heat input from the stirrer, is given by Eq. 8.1. The heat accumulated is equal to the heat produced by the reaction minus the heat dissipated by the external cooling system:

$$\dot{q}_{acc} = \dot{q}_r - \dot{q}_c \tag{8.1}$$

- $\dot{q}_{acc}$: rate of heat accumulation
- $\dot{q}_r$: rate of heat production (reaction heat)
- $\dot{q}_c$: rate of heat removal (cooling)

8.2.2   Heat Production

The *heat production* term ($\dot{q}_r$) in Eq. 8.1 corresponds to the rate of heat generated by the chemical reaction. Heat production[2] is thus proportional to the reaction rate and the reaction enthalpy as shown in Eq. 8.2:

$$\dot{q}_r = \frac{1}{\rho} (-r_A)(-\Delta H_R) \qquad (8.2)$$

- $\dot{q}_r$: rate of heat production [W/kg]
- ρ: density of the reaction mass [kg/m^3]
- r_A: reaction rate with respect to reactant A [mol m^{-3} s^{-1}]
- ΔH_R: reaction enthalpy [J/mol]
- Assumption: ρ and ΔH_R are temperature independent

The reaction rate r_A is temperature and concentration dependent and can be defined by Eq. 8.3:

$$r_A = k(T) \times f(C_A) \qquad (8.3)$$

- k: temperature-dependent reaction rate constant
- $f(C_A)$: function that depends on the concentration and the conversion of reactant A [mol/m^3]

The reaction rate constant k is defined by the Arrhenius equation:

$$k = A e^{\frac{-E_A}{RT}} \qquad (8.4)$$

- A: pre-exponential factor [1/s] (unit for first-order reaction rate constant)
- E_A: activation energy [J/mol]
- R: universal gas constant [J mol^{-1} K^{-1}]
- T: reaction temperature [K]

Notice that the reaction rate constant increases exponentially with temperature, leading to an exponential increase in the reaction rate and hence in the production of heat.

The heat produced by a chemical reaction is also proportional to the enthalpy change of the reaction (ΔH_R), which makes it an important indicator for thermal hazards. Standard values of the reaction enthalpy for common synthesis and decomposition reactions in the chemical industry are shown in Table 8.1.

---

[2]Note that in the context of thermal process safety, all effects that increase the temperature are positive (e.g., exothermic reactions).

Table 8.1 Standard values of the enthalpy of reaction $(-\Delta H_R)$ for common synthesis and decomposition reactions in the chemical industry. Adopted from Stoessel (1993) and Gygax (1993)

Reaction	$-\Delta H_R$ [kJ/mol]
Neutralization (HCl)	55
Diazotization	65
Amination	120
Nitration	130
Decomposition (diazo)	140
Sulfonation	150
Polymerization (vinyl)	50–200
Decomposition (nitro)	400
Hydrogenation (nitro)	560

Decomposition reactions occur when a substance or reaction mixture moves from a thermally unstable state to a more stable one. These reactions often produce a large amount of thermal energy and can lead to a thermal runaway. They can occur unintentionally during synthesis reactions when the generated heat is not appropriately removed, in storage vessels, or during other processes (e.g., during the grinding of cyanuric chloride). The presence of specific functional groups in a molecule can also play an important role in promoting decomposition reactions. Examples of typical functional groups present in unstable compounds with high decomposition potential are shown in Table 8.2.

Table 8.2 Typical functional groups of unstable compounds

Perchlorate	Nitro	Polynitro	Nitroso	N-oxide
$-ClO_4$	$-NO_2$		$-N=O$	$\diagdown N^+ - O^-$ $\diagup$
Hydroxylamine, oxime	Tetrazole	Triazene, triazole	Azo	Hydrazine
$>N-OH$	(tetrazole ring structure)	$-N=N-N<$	$-N=N-$	$-NH-NH_2$, $>N-NH_2$
Substituted hydrazine	$-N-N-$ in a ring	2 $-N-N-$ in a ring	Imidazole	Oxazole
$-NH-NH-$, $>N-N<$			$-N=C-N-$ (ring)	$-N=C-O-$ (ring)
Thiazole	Acetylide	Halogen/nitrogen compounds	Nitric acid esters	Peroxides
$-N=C-S-$(ring)	$-C\equiv C-$	$>N-X$	$-ONO_2$	$-O-O-$

8.2.3   Heat Removal

The *heat removal* term ($\dot{q}_c$) in Eq. 8.1 corresponds to the cooling provided by a system installed within the chemical process. For a batch reactor, this could be, for example, through heat transfer to an external cooling jacket[3] as illustrated in Fig. 8.1. This rate can be expressed as:

$$\dot{q}_c = \frac{U}{\rho} \frac{A}{V} (T - T_c) \qquad (8.5)$$

- $\dot{q}_c$: rate of heat removal by heat transfer [W/kg]
- T: temperature of the reaction mass [K]
- T_c: coolant temperature [K]
- A: cooling area [m^2]
- V: reaction volume [m^3]
- U: overall heat transfer coefficient [W m^{-2} K^{-1}]
- Assumptions: U is temperature independent, the temperature of the coolant is constant, and there is neither an axial nor radial concentration or temperature gradient in the reactor content

It is important to see that the heat removal varies linearly with temperature, and therefore the heat removal rate increases more slowly than the heat production rate (which increases exponentially). When scaling up a process, it is, therefore, necessary to consider that the specific cooling area (A/V) decreases in proportion to the scale-up factor. Using larger vessels can limit the conditions needed for proper heat removal. In addition, resistance to heat transfer can stem from the poor heat conduction of larger solid masses, and quasi-adiabatic[4] conditions can often apply to the center of a container. Without forced cooling, even the smallest heat production rates of a decomposition reaction (e.g., a few W/m^3) can lead to heat accumulation within the system. This can occur in the case of both chemical storage and also in a batch reactor with a stirrer failure.

During a cooling system failure, the values of the heat transfer coefficients are approximately an order of magnitude lower than during normal operating conditions.

8.2.4   Heat Accumulation

The *heat accumulation* ($\dot{q}_{acc}$) calculated in Eq. 8.1 represents the variation of energy in the system with temperature. It can also be defined for a batch reactor using the

---

[3]Note that other forms of heat removal are possible, for example, through evaporation or melting processes (e.g., using solvent or ice).

[4]In an adiabatic process, no heat is transferred to or from the environment surrounding the system.

specific heat capacity and the change in temperature as:

$$\dot{q}_{acc} = c_p \frac{dT}{dt} \qquad (8.6)$$

- $\dot{q}_{acc}$: rate of heat accumulation [W/kg]
- c_p: specific heat capacity of the reaction mixture [J kg^{-1} K^{-1}]
- T: temperature of the reaction mass [K]
- t: time [s]
- Assumptions: c_p is temperature independent, and the heat capacity of the equipment can be neglected[5]

The specific heat capacity of a mixture ($c_{p_{mix}}$) can be estimated from the mass (m) and specific heat capacities (c_p) of the individual components (i) in the mixture as:

$$c_{p_{mix}} = \frac{\sum_i (m_i \times c_{p_i})}{\sum_i m_i} \qquad (8.7)$$

8.2.5   Semenov Diagram and the Critical Temperature

Substituting each of the these three definitions for $\dot{q}_{acc}$, $\dot{q}_r$, and $\dot{q}_c$ in Eq. 8.1, the simplified heat balance for an exothermic batch reaction can be expressed as:

$$\underbrace{c_p \frac{dT}{dt}}_{\substack{\dot{q}_{acc} \\ \text{accumulation}}} = \underbrace{\frac{1}{\rho} (-r_A)(-\Delta H_R)}_{\substack{\dot{q}_r \\ \text{reaction}}} - \underbrace{\frac{U}{\rho} \frac{A}{V} (T - T_c)}_{\substack{\dot{q}_c \\ \text{external cooling}}} \qquad (8.8)$$

A helpful way to describe and analyze the heat balance and a reactor's resulting stability during normal operating conditions is by plotting the heat balance[6] on a Semenov diagram, as shown in Fig. 8.2. In this diagram, the heat produced by the reaction ($\dot{q}_r$) and the heat removed by the cooling system ($\dot{q}_c$) are plotted against reaction temperature (T). As shown, an increase in reaction temperature, according to the Arrhenius relationship in Eq. 8.4, is associated with an exponential increase in the reaction rate and hence in the heat produced. In contrast, heat removal by an external cooling medium only increases linearly with reaction temperature.

The slope of the heat removal line ($\dot{q}_c$) is proportional to the overall heat transfer coefficient (U) and the ratio of the cooling area (A) to the reaction volume (V).

---

[5]For stirred tank reactors, such as batch reactors, the heat capacity of the reactor is often negligible compared to the heat capacity of the reaction mass.

[6]This heat balance considers zero-order kinetics, which is a conservative approximation used in the context of thermal process safety.

Fig. 8.2 Heat balance of a batch reaction in the form of a Semenov diagram. T_c: coolant temperature, $T_{c, critical}$, critical temperature of the coolant; $T_{critical}$, critical reaction temperature where the slope of the heat produced is equal to the slope of the heat removed; $\dot{q}_r$, heat produced by the reaction; $\dot{q}_c$: heat removed by the cooling system

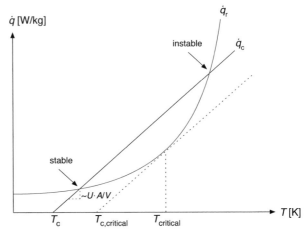

The intersection with the abscissa represents the coolant temperature (T_c). Heat production is equal to heat removal ($\dot{q}_r = \dot{q}_c$) at the two points where the plotted lines intersect, representing that the heat balance is in equilibrium.

The stable operating reaction temperature corresponds to the temperature at the lower intersection point. At a reaction temperature above the temperature corresponding to the upper intersection point, the reaction mass cannot be properly cooled. If the coolant temperature (T_c) is increased, the heat removal line ($\dot{q}_c$) shifts in parallel to the right as shown by the dotted line in Fig. 8.2. In the case that the coolant temperature reaches the critical point ($T_{c, critical}$), the two intersection points move together until they meet at the tangent point known as the critical reaction temperature ($T_{critical}$). If the temperature of the cooling system is increased above this critical cooling temperature ($T_{c, critical}$), the lines no longer intersect, which means the heat balance equation has no steady-state solution and a thermal runaway reaction will ultimately begin.

Operating the reactor close to this critical cooling temperature means that even a small change in the overall heat transfer coefficient (U), the heat transfer area (A), or the temperature of the coolant (T_c) could lead to a thermal runaway.

Taking into account that at the critical temperature the slope of the rate of heat produced and the slope of the rate of heat removed are equal, the following expression is obtained, which gives an indication of the reactor's thermal stability:

$$T_{critical} - T_c > \frac{R \times T_{critical}^2}{E_A} \tag{8.9}$$

The greater $T_{critical} - T_c$ is, the more thermally stable the reactor is.

8.3 Assessing Thermal Risks

Having introduced the fundamental concept of heat balances for ensuring reactor stability, this section goes on to present some of the key indicators used to assess thermal risks within a process. It also provides an overview of some of the basic guidelines used to estimate the impact and probability of a thermal runaway reaction as well as methods to evaluate and manage identified thermal risks.

8.3.1 Thermokinetic Concepts

Fundamental to understanding the thermal hazards within a system are the concepts of adiabatic temperature rise (ΔT_{ad}), the maximum temperature of the synthesis reaction ($MTSR$), and the time to maximum rate under adiabatic conditions (TMR_{ad}). The calculations behind each of these are briefly introduced here, and they will be further examined and applied later in the chapter.

8.3.1.1 Adiabatic Temperature Rise

The *adiabatic temperature rise* (ΔT_{ad}) is the increase in temperature in the system without heat exchange with the surroundings (i.e., under adiabatic conditions). Considering this, it can be derived from the heat balance defined in Eq. 8.8 to be:

$$\Delta T_{ad} = \frac{Q_r}{c_p} = \frac{-\Delta H_R \times C_0}{\rho \times c_p} \tag{8.10}$$

- ΔT_{ad}: adiabatic temperature rise [K]
- Q_r: heat produced by chemical reaction [J/kg]
- c_p: specific heat capacity of the reaction mass [$J\,kg^{-1}\,K^{-1}$]
- ΔH_R: reaction enthalpy [J/mol]
- C_0: initial reactant concentration [mol/m^3]
- ρ: density of the reaction mass [kg/m^3]

The experimental determination of this key indicator for thermal risk is briefly introduced in Appendix F.

8.3.1.2 Maximum Temperature of the Synthesis Reaction

The *maximum temperature of the synthesis reaction (MTSR)* is the maximum temperature that can be reached by the synthesis reaction under adiabatic conditions. This is dependent on the process temperature, the accumulation of reactant, and the adiabatic temperature rise. For batch reactions, the accumulation is greatest at the beginning of the reaction. In this case, the *MTSR* can be expressed according to Eq. 8.11:

$$MTSR \cong T_{process} + \Delta T_{ad} \tag{8.11}$$

- *MTSR*: maximum temperature of the synthesis reaction [K]
- $T_{process}$: temperature of the desired reaction process [K]
- ΔT_{ad}: adiabatic temperature rise [K]

The *MTSR* is used during the development of the process design as a basis to ensure reactor stability even in case of malfunctioning (e.g., cooling failure).

8.3.1.3 Time to Maximum Rate Under Adiabatic Conditions

The *time to maximum rate under adiabatic conditions* (TMR_{ad}) is the amount of time needed before a thermal explosion takes place under adiabatic conditions. It can be calculated solving the heat balance for a zero-order[7] reaction under adiabatic conditions. This can be done since for large values of ΔT_{ad}, the Arrhenius term of the reaction rate becomes much more important than the concentration-dependent term. The TMR_{ad} [s] is then defined as:

$$TMR_{ad} \cong \frac{c_p}{\dot{q}(T_0)} \frac{RT_0^2}{E_A} \tag{8.12}$$

- c_p: specific heat capacity of the reaction mixture [J kg^{-1} K^{-1}]
- T_0: temperature of the reaction at the initial conditions [K]
- E_A: activation energy [J/mol]
- R: universal gas constant [J mol^{-1} K^{-1}]
- $\dot{q}(T_0)$: rate of heat produced by the chemical reaction at temperature T_0 [W/kg]

The TMR_{ad} can be calculated to represent both (1) how quickly the runaway of the desired reaction occurs and (2) how fast the runaway of a decomposition reaction occurs starting at the *MTSR*. In the first case, the T_0 will be equal to the operating temperature of the process, and in the second case, the T_0 will be equal to the *MTSR*.

To calculate TMR_{ad} according to Eq. 8.12, the rate of heat produced ($\dot{q}$ at $T = T_0$), and the activation energy (E_A) of the reaction must be known. Methods to experimentally determine these parameters are briefly introduced in Appendix F.

8.3.2   Cooling Failure Scenario

The scenario of a cooling system failure during an exothermic batch reaction has proven to be a useful case for learning about the thermal risks of chemical processes. In order to conservatively identify its risks, the following assumptions can be made:

---

[7]Even though the concept of TMR_{ad} was developed for zero-order kinetics, it still provides a conservative approximation for strongly exothermic reactions with higher kinetic orders (Stoessel, 2008).

- Adiabatic reaction conditions: upon cooling system failure, the heat transfer is severely limited. As an adiabatic system is independent of technical operating parameters, it can, therefore, be easily represented in a model.
- Batch reaction: unconverted reactants are present at the time of the cooling failure.

Figure 8.3 illustrates the evolution of the temperature in the reactor following a cooling failure in an exothermic batch reaction. A standard batch reaction involves loading the reactants into the reactor vessel at room temperature and heating while stirring the mixture until reaching the process reaction temperature ($T_{process}$). The temperature is then held constant during the reaction time. After completion of the reaction, the reactor is cooled and emptied.

If a cooling failure occurs when the reaction mixture is at $T_{process}$, and if unconverted reactants are still present in the reactor, then under adiabatic conditions, the temperature will increase due to the production of reaction heat until the *MTSR* is reached. The exact value of the *MTSR* depends on the amount of non-reacted material present, and it may be high enough to initiate a decomposition reaction. The heat produced during the decomposition reaction can then further increase the system temperature until a thermal explosion occurs and an end temperature (T_{end}) is reached. As shown in Fig. 8.3, both the desired reaction and the decomposition reaction are connected through the *MTSR* in this scenario and can be simultaneously analyzed.

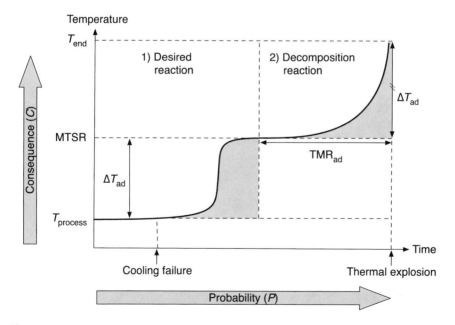

Fig. 8.3 Cooling failure scenario during an exothermic batch reaction

The key questions to answer here when assessing the risks related to a runaway scenario can be formulated in terms of temperature and time as follows:

(i) Temperature:

- What process operating temperature ($T_{process}$) allows for the safe dissipation of the heat of reaction during normal operation?
- What is the maximum temperature achieved by the runaway of the synthesis reaction ($MTSR$) in the event of a cooling breakdown?
- Which final temperature (T_{end}) could be reached following a possible subsequent decomposition reaction?

(ii) Time:

- At what point in time would a failure of the cooling system be most critical?
- How much time does a runaway of the desired chemical reaction need? (However, this time frame is usually short and therefore does not provide a safety factor.)
- How much time does the runaway of the decomposition reaction need? (TMR_{ad} represents the time available to take corrective measures.)

Identifying the ΔT_{ad}, $MTSR$, and TMR_{ad} helps to answer these questions and ensures that the thermal risks for the chemical process have been identified.

Subsequently, these values can also be used to characterize the risk of a runaway reaction by its consequences and probability of occurrence.

8.3.3 Consequences of a Thermal Runaway

The adiabatic temperature rise (ΔT_{ad}) defined in Eq. 8.10 can be used as an indicator for the potential consequences of a thermal runaway reaction. Table 8.3 shows classifications of ΔT_{ad} that can be used as a descriptor of the potential impact of a thermal runaway.

Table 8.3 Classification of the impact of a thermal runaway using the value of adiabatic temperature rise (Stoessel, 2008)

ΔT_{ad} [K]	Impact
>200	High
$50 < \Delta T_{ad} < 200$	Medium
$<50^{a}$	Low

[a]Boiling point not exceeded as an additional condition

8.3.4   Probability of a Thermal Runaway

For reactions with medium and high thermal impact potentials ($\Delta T_{ad} > 50\,K$), the probability of occurrence should also be analyzed. The time to maximum rate under adiabatic conditions (TMR_{ad}) according to Eq. 8.12 can be used as an indicator for the probability of a thermal runaway. Table 8.4 shows classifications of TMR_{ad} that can be used to help describe this probability.

Table 8.4 Classification of TMR_{ad} as a descriptor for the probability of occurrence of a thermal runaway following a cooling system failure (Stoessel, 2008)

TMR_{ad} [hr]	Probability
<8	High
$8 < TMR_{ad} < 24$	Medium
>24	Low

The maximum temperature of the synthesis reaction ($MTSR$) is usually reached very quickly after a cooling failure. It is therefore crucial to the success of any implemented safety measures that the TMR_{ad} is sufficiently high and, as shown in Table 8.4, preferably greater than 24 hr. A safe $MTSR$ (equal to T_0)[8] can, therefore, be calculated with Eq. 8.12 by setting TMR_{ad} equal to 24 hr and using the experimentally determined values for the thermal decomposition power and the activation energy (see Appendix F). Safety measures should be implemented for reactions classified with a medium or a high probability.

8.3.5   Risk Evaluation

As presented in Chap. 7, after the consequences and the probability of an accident scenario have been estimated during a process risk assessment, they are then combined to determine the level of risk. Based on this risk level and set acceptability criteria, the evaluation step of process risk assessment then determines which risks are acceptable and which need to be reduced. The same is true for the risk evaluation of thermal hazards.

One common way to estimate and communicate the level of risk is through a risk matrix, and this technique can also be applied to thermal hazards. An additional technique is through the definition of *criticality classes* based on temperature levels.

8.3.5.1 Risk Matrices

A risk matrix helps determine and visualize a level of risk, by combining scales of increasing probability on one axis and consequence severity on the other. For thermal hazards, the adiabatic temperature rise is used as a proxy measure for the consequence on one axis, and the time to maximum rate under adiabatic conditions as a proxy measure for the probability on the other.

---

[8]Indicated in Fig. 8.5 as T_{24}.

Figure 8.4 illustrates such a risk matrix for thermal hazards. Unacceptable risks are determined by setting an acceptability line within the matrix. Risks above this line are considered to be not acceptable and require risk management.

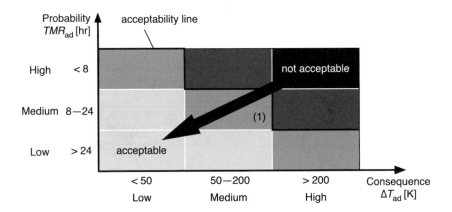

Fig. 8.4 Example of a risk matrix for thermal hazards. 1) Taking additional safety measures is advised in this transition area

8.3.5.2 Criticality Classes

Another way to evaluate the thermal risk of a cooling failure scenario during an exothermic reaction is through the use of *criticality classes*. These classes are based on the order in which the following four characteristic temperature levels are reached following the cooling failure (Stoessel, 2008):

- $T_{process}$: the process temperature
- *MTSR*: the maximum temperature that can be reached by the synthesis reaction under adiabatic conditions
- T_{24}: the temperature at which TMR_{ad} is 24 hr. This is the highest temperature at which the thermal stability of the reaction mass is not problematic (according to Table 8.4).
- *MTT*: the maximum tolerable temperature for a system. For open systems, this is the boiling point, and for closed systems, this is the temperature at the maximum allowable pressure (i.e., before bursting the safety valve or activating the rupture disk).

Five different criticality classes exist using these four temperature levels. The order in which the temperatures are reached for each class is illustrated in Fig. 8.5. The reasoning for their criticality is given as follows:

- *Criticality class 1*: Reactions of this class are not critical since the *MTT* cannot be reached and the decomposition reaction cannot be triggered as the value of

MTSR stays below the temperature required to reach T_{24}. These reactions can be considered inherently safe.

- *Criticality class 2*: This type of reaction is also not critical; however, the *MTT* of the system is no longer available to serve as a safety buffer to avoid a decomposition reaction.
- *Criticality class 3*: In these reactions, the *MTSR* exceeds the *MTT*; however a decomposition reaction cannot be triggered. Careful attention should be paid to ensure a sufficient design of the evaporative cooling system or depressurization control. However, this reaction scenario is otherwise not critical.
- *Criticality class 4*: Within this class, the thermal potential of the desired reaction exceeds the T_{24}, and a thermal decomposition could occur. The lower *MTT* value of the system can at best help serve as a safety buffer to stabilize the temperature, but careful attention should be paid to a sufficient design of the condensation system. In any case, this scenario is critical, and a technical safety measure is required to manage the risk.
- *Criticality class 5*: In this highest class, the thermal potential of the desired reaction exceeds the T_{24}, and a thermal decomposition could occur without any safety buffer provided by the *MTT*. Such reactions are critical and should be shifted to a lower criticality class through a redesign of the reaction process.

Inherent process safety decreases from criticality class 1 (highest inherent safety) to criticality class 5 (no inherent safety).

8.3.6   Risk Management

For risks determined to be not acceptable, *risk management* actions should be taken to reduce their level of risk. The principles of inherent process safety and functional process safety (layers of protection) introduced in Chap. 7 are also applicable to the risk reduction of thermal hazards. In decreasing order of importance, some of the measures that can be taken include (Kletz and Amyotte, 2010):

Inherently safer design:

- Minimization of material and energy content in the process (e.g., use of a continuous process instead of a batch process)
- Substitution of the reactants and solvents (e.g., with non-hazardous substances, ensure no unstable intermediates are generated, no highly energetic compounds used).
- Use of less hazardous process conditions (e.g., lower temperatures,[9] dilution,[10] etc.).

---

[9]Attention: this could result in a high accumulation of unreacted reactants.

[10]Note that a side effect of diluting the reaction mixture could be adverse environmental and economic impacts.

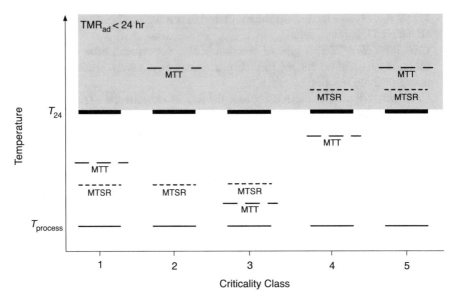

Fig. 8.5 Criticality classes to support the risk evaluation of a cooling failure scenario during an exothermic reaction. These classes are defined according to the order reached by the process temperature (T_p), the maximum temperature that can be reached by the desired reaction under adiabatic conditions (*MTSR*), the temperature at which TMR_{ad} is 24 hr (T_{24}), and the maximum tolerable temperature (*MTT*). Adapted from Stoessel (2008)

- Use of a reactor design with pressure resistance high enough to resist the maximum pressure generated from a worst-case runaway reaction (eliminates the need for a large emergency pressure relief system).
- Use of dedicated reaction equipment for each process step.

Preventative measures:

- Control the reaction feed to limit the accumulation of non-converted reactants. For example, by using a semi-batch process where the feed rate and feed temperature are additional process parameters. This provides more options for controlling the reaction.
- Control the reaction temperature by maintaining the intended heating rates.
- Control the loading of reactants to ensure the defined amount and quality is used.
- Implement an emergency cooling system.
- Implement an emergency system that can slow down or stop a reaction through quenching (e.g., adding an agent that can interrupt a catalyst or modify the pH) and flooding (e.g., adding a large amount of an inert material that can dilute and cool down the reaction mass).

Mitigative measures:

- Implement a system that allows for a controlled depressurization of the reactor at the early stages of a thermal runaway.
- Install an emergency pressure relief system that allows gases and vapor to escape following a rapid increase in pressure.
- Implement an emergency containment system that keeps the contents from the reactor in a controlled area after a thermal runaway.

8.4   Example: Wet Grinding of Cyanuric Chloride

In Chap. 7, the grinding process of cyanuric chloride was first introduced as an example to illustrate the general procedure for the risk assessment of chemical processes. As part of this example, the thermal hazards associated with the decomposition reaction of cyanuric chloride also have to be assessed.

In this section, this will be done by estimating the impact and probability of a thermal runaway from this decomposition reaction. For a description of the process and the hydrolysis decomposition reaction itself, see Chap. 7.

8.4.1   Thermal Consequence Analysis

To assess the potential consequences of a thermal runaway from the decomposition reaction, a worst-case scenario is assumed, namely, adiabatic conditions. Using the following reaction data and Eq. 8.10, the adiabatic temperature rise (ΔT_{ad}) for this reaction system can be calculated:

- Total mass of the reaction mixture: 698 kg
- Amount of cyanuric chloride in the reaction mixture: 1.05 kmol
- Enthalpy of reaction: −450 kJ/mol cyanuric chloride
- Specific heat capacity of the reaction mass: 3.6 kJ kg^{-1} K^{-1}

$$\Delta T_{ad} = 450\frac{kJ}{mol} \times \frac{1050\,mol}{698\,kg} \times \frac{1\,kg \times K}{3.6\,kJ} = 188\,K$$

According to the criteria defined in Table 8.3, this reaction has a medium potential impact. Even if the enthalpy of fusion of ice (83,000 kJ for 250 kg of ice) used for cooling is subtracted from the total heat produced by the decomposition reaction, there is still an adiabatic temperature rise of 156 K and the boiling point of water as the wet grinding solvent is also reached (still considered as a medium potential). Therefore, the grinding process of cyanuric chloride cannot be considered inherently safe regarding thermal hazards.

8.4.2   Thermal Probability Analysis

Considering this heightened level of potential consequence, it is necessary to estimate the probability of a runaway reaction occurring. To do this, the time to maximum rate under adiabatic conditions (TMR_{ad}) can be calculated. For this calculation, additional kinetic data is needed including the rate of heat release by the hydrolysis reaction, activation energy, and pre-exponential factor (see Eq. 8.4).

For this specific example, this information is not available. However, from experimental studies reported in the literature (Yan et al., 2008), it is known that cyanuric chloride is stable (hydrolysis rate very slow) for approximately 12 hr between 0 and 2 °C and at pH 7.

Considering this and the classifications presented in Table 8.4, the probability of a runaway reaction for the grinding process can be classified as medium.

8.4.3   Thermal Risk Evaluation

Combining these results into the risk matrix presented in Sect. 8.3.5, Fig. 8.6 shows the resulting acceptability of the risk posed.

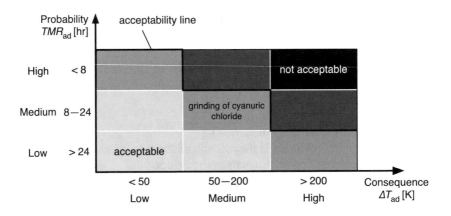

Fig. 8.6 Risk matrix for thermal hazards in the grinding process of cyanuric chloride

This matrix indicates that the thermal risk related to this process is just below the acceptability line. It is crucial to ensure control of the temperature and pH during the grinding process as well as implementation of a reliable depressurization and ammonia absorption system (as mitigative safeguards). This outcome is used as input into the full process risk assessment for the wet grinding of cyanuric chloride included in Chap. 7 (see Sect. 7.9).

References

Gygax R (1993) Thermal Process Safety. ESCIS Safety Series, 8. https://doi.org/10.3929/ETHZ-B-000354503

Kletz T, Amyotte P (2010) Process Plants: A Handbook for Inherently Safer Design, 2nd edn. CRC Press. https://www.crcpress.com/Process-Plants-A-Handbook-for-Inherently-Safer-Design-Second-Edition/Kletz-Amyotte/p/book/9781439804551

Stoessel F (1993) What is your thermal risk? Chemical Engineering Progress 89(10):68–75

Stoessel F (2008) Thermal Safety of Chemical Processes. Wiley-VCH Verlag GmbH & Co. KGaA, Weinheim, Germany. http://doi.wiley.com/10.1002/9783527621606

Yan Z, Xue WL, Zeng ZX, Gu MR (2008) Kinetics of Cyanuric Chloride Hydrolysis in Aqueous Solution. Industrial & Engineering Chemistry Research 47(15):5318–5322. https://pubs.acs.org/doi/10.1021/ie071289x

Societal Dialogue on Risks and Benefits

9

9.1 Risk: A Balancing Act for Society

Whether natural or man-made, *risks* have always existed and posed a threat to human life on Earth. However, the acceptance of some risks has allowed for *benefits* in the form of exciting innovations that have improved quality of life and launched new technologies previously unthinkable. Over the past century, the rapid and progressive development of technology has led to a shift toward more and more man-made risks (see Chap. 2). Many of these risks are now associated with systems and technologies that have become so complex that they are understandable only to technical experts and even then with increasing uncertainty. This is in addition to the natural complexity of human health and the environment, which can require complicated (and even non-existent) methods to properly model and test for adverse effects. To complicate this further, rapid globalization has made weighing the risks and benefits of chemical technology a very international affair. Society has also been shifting toward increased individualism leading to a wide range of lifestyles, preferences, and worldviews. All of these aspects have created the need for more and more international *dialogue* with an increasing number of stakeholders, each having their own varied, coexisting values and protection goals.

Achieving prosperity without taking on risk seems to be inherently impossible, and recent developments of technology underline this. However, at a certain point, accepting a risk might be considered too perilous for the prosperity it may provide. From this point of view, progress itself could at times be seen as just as risky as a lack of progress. Today, more than ever, a fundamental question to answer is: Given the expected benefits of a chemical technology, what risk is society willing to take?

The earlier chapters on life cycle assessment (LCA), product and process risk assessment, and thermal safety introduced methods and tools to help to identify, understand, and manage the various types of risks that are associated with chemical products and processes, specifically in regard to the guiding principles *eco-efficiency* and *inherent safety*. This chapter introduces *risk-benefit dialogue* as a method to support the third and final guiding principle of integrated development: *social acceptability*. Risk-benefit dialogue aims to do, quite simply, exactly what it sounds like—promote dialogue discussing the risks and benefits of a chemical technology among the stakeholders who are potentially affected by it.

Even with the best safety experts involved in the development of a product or process, it should never be assumed that the larger stakeholder community would not be able to understand and contribute to determining the acceptability of a risk. Societal dialogue should be implemented to support society's understanding of a risk and help stakeholders view relevant protection goals and technology as a shared responsibility. Gaining the public's trust and acceptance of a product or process early on can address existing or potential future concerns that could ruin a company's reputation. Achieving this, however, can be a complex task requiring significant effort, especially for new products that society is not yet familiar with. It demands a dedication to transparency and the willpower to bring stakeholders into the discussion and keep them involved.

The risks and benefits that need to be considered in the dialogue are often not hard facts. Purely technical risk calculations that often come to mind when talking about risk (e.g., based on probability and impact) are just one way of presenting risks. Since risks are difficult for humans to rationally perceive, subjective differences in the definition and assessment of risks will also always exist across stakeholders (see Chap. 2). This is exactly where the risk-benefit dialogue plays an important role.

In the past, societal dialogue has largely focused on discussing short-term, *acute* risks such as chemical factory accidents (see a list of examples in Appendix A). These highly visible events led to noticeable, localized damage to human and environmental health. Recently, topics such as climate change and widespread marine plastic pollution have also become increasingly visible, highly discussed topics. Concern from stakeholders has created increasing pressure for regulators and industry to make changes to address them.

However, unlike factory accidents and marine plastic litter, the presence of hazardous chemicals in drinking water, consumer products, or food is usually not visible at all to the naked eye. At the same time, this chemical pollution may also pose long-term, *chronic* risks with damage that might not be discovered until decades after they begin. Encouraging stakeholders to discuss risks they cannot see and may not feel the full consequences of for many years is a more difficult but still very important undertaking. Ongoing research into the potential scale of adverse effects on human health from chronic chemical exposure highlights that ignoring these less visible risks could have significant, global consequences.

This chapter introduces the risk-benefit dialogue as an important method for both the chemical industry and wider society. It discusses the diverse stakeholders that can be involved, introduces ways to consider the variety of viewpoints that can exist, and provides guidance on communicating risks and managing risk-benefit dialogues depending on the case for its use. Further, in-depth discussion and guidance on facilitating and promoting societal dialogue can be found in key literature from several discussion papers and project outcomes cited throughout the chapter.

9.2   Diverse Viewpoints

The interactions between the chemical industry and the rest of society can be extremely diverse. Figure 9.1 depicts the variety of stakeholders that can interact with and be affected by the industry's operations and actions. Understanding the many viewpoints, concerns, and interests of these stakeholder groups is important to ensuring the societal acceptance of the industry and also for the industry to generate better products and processes.

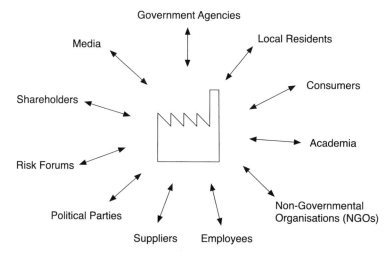

Fig. 9.1 Some of the important societal stakeholders for the chemical industry

In a commentary in the journal *Nature Chemistry* (Hartings and Fahy, 2011), authors Matthew Hartings and Declan Fahy highlight that the public often suffers from a developed fear of chemistry derived from its fundamentally complex nature, as well as negative historical associations, brought about by events such as chemical warfare in the First World War and industrial accidents. Recent international topics such as plastic pollution, global warming, and drinking water pollution from lead pipes and PFAS are not helping to mitigate this. Low levels of public confidence in the chemical industry can likely be linked to a combination of several factors:

- Dramatic chemical accidents
- Suspected as well as manifest health and environmental damage from chemical products or process emissions
- Limited knowledge of the chemical industry and its products and processes
- Lack of communication between the chemical industry and societal stakeholders

Perhaps an important, ongoing development affecting the perspective of the general public is the rise of *citizen scientists* and public information platforms hosting scientific information, such as Wikipedia. The advancement of technology has allowed for the layperson to generate and share scientific data using open-source software and affordable measurement devices, and the growth of the internet and shift toward open-access platforms has also placed new data in the hands of the public. While there are some concerns about the accuracy and reliability of citizen scientists (Nature, 2015), they have nevertheless become informed stakeholders in risk-benefit discussions.

Examples of some of the most notable recent citizen science initiatives are directly related to incidents of pollution caused by man-made technology, such as local sampling of lead levels in drinking water following the widespread lead contamination in Flint, Michigan, US (Flint Water Study, 2016), and measuring long-term radiation levels in Japan following the Fukushima nuclear power plant disaster (Beser, 2016). Review papers have been written on examples within other fields (Wylie et al., 2017b,a; Brook and McLachlan, 2008), and the online German platform *Bürger schaffen Wissen* (loosely translated as "Citizens Create Knowledge") (Bürger schaffen Wissen, 2019) is an example of a wider initiative to promote citizen science and share results. Social media platforms have also become important outlets for citizen scientists to share results and for citizens in general to express their views of controversial technologies. Public data from social platforms are even being used to examine and better understand these trends in public opinion (Müller et al., 2019).

To support dialogue among such a diverse set of stakeholders, some neutral, formally organized forums have been created for discussing the risks and benefits of various technologies within society. In Switzerland, the Risk Dialogue Foundation was founded in 1989 following highly visible accidents in the 1980s such as the Chernobyl nuclear disaster, the Schweizerhalle chemical spill in Basel, and the Challenger space shuttle explosion (Stiftung Risiko-Dialog, 2019). Since then it has initiated and moderated several public discussions on topics such as energy and network risks, nanotechnology, and genetic engineering. The Swiss government has also organized public forums on topics such as nanotechnology (Swiss Federal Office for Public Health, 2018).

In Germany, the organization *Wissenschaft im Dialog* (Science in Dialogue) was founded in 1999 to improve the communication of science with the public (Wissenschaft im Dialog, 2019). One of its activities is the creation of a discussion platform called *Die Debatte* (The Debate) that organizes and hosts discussions with various stakeholders on topics such as plastic waste, the collapse of bee populations, and fine particulate air pollution (die Debatte, 2019). Through freely accessible recorded debates, short publications, and public outreach, the platform aims to bring scientific facts and perspectives into societal discussions on key issues.

9.3 Aims and Levels of Societal Dialogue

Following a technical risk assessment (see Chaps. 6, 7, and 8), *residual risks* in particular need to be addressed through a risk-benefit dialogue with relevant stakeholders. Overall, a risk-benefit dialogue should seek to:

- Find risk evaluation criteria that are as widely supported as possible
- Allow for the participation of skeptical stakeholders or of those with opposing views
- Gain understanding and acceptance from an informed public
- Understand the reasons for any critical, subjective assessments of risks

A risk-benefit dialogue can be understood as comprising three levels:

- The *technical level*: Includes the technical risk calculation and development of technical management options to minimize residual risk. Efforts are led by technical experts.
- The *perception level*: Identifies assessment and measurement criteria for the risks. Efforts are led by natural and social science experts.
- The *decision-making level*: Decides on the acceptability of the risk. These decisions are often made within governmental institutions such as public health, safety, and environmental authorities.

On the *technical level*, risk is primarily centered around technical issues relating to plants, processes, and products. The business environment involving customers and suppliers also plays an important role. Through risk assessment, potentially critical conditions are determined through data collection and the creation of scenarios resulting in estimated impacts and their probabilities of occurrence. Once specific, protective measures have been defined to manage the risk, the residual risk still needs to be determined.

On the *perception level*, the focus is on the perception and assessment of the risks and benefits by particular individuals and stakeholders, who often have diverse perspectives regarding a technology. In a society with a multitude of lifestyles, value systems, and world views, differentiated groups emerge that have different patterns of perception and logical reasoning. In particular, differences in perception can often arise when a technology benefits other stakeholders than those that would be subjected to the associated risk(s). Such stakeholders may have little to no basic knowledge about the scientific or technical aspects of the technology.

The *decision-making level* of risk-benefit dialogue then considers the residual risk determined on the technical level as well as the evaluation criteria on the perceptive level to engage in a dialogue with all stakeholders and arrive at a decision on whether the risk is acceptable or not. This is often facilitated and/or managed within the legal and regulatory framework(s) in place, and it results in updated regulations that reflect the level of risk deemed acceptable.

9.3.1 Considering Risk Aversion

Many individuals dislike accidents with high impact potential, even if the probability of occurrence is very low. To help with assessing this *risk aversion* on both the technical and perceptive levels, it can be considered, for example, by introducing an aversion factor into the product formula for calculating risk:

$$R = p \times I^{\alpha} \tag{9.1}$$

- *R*: risk
- *p*: probability
- *I*: impact
- α: empirically determined *risk aversion factor* ($\alpha > 1$, i.e., risks that are less likely but of high impact are weighted more heavily)

In considering a risk and preparing for risk communication with stakeholders, this calculation can help to consider the subjective societal perception of risk alongside the technically-based criteria. A greater α means a steeper acceptability line in the risk diagram discussed in the following section and presented in Fig. 9.2.

9.3.2 Using Risk Diagrams

A helpful way to illustrate protection goals and map the limits of societal acceptance (as defined on the decision-making level) to support dialogue is through the use of a *risk diagram*. This is very similar to the use of a risk matrix as introduced in Chap. 7. An example of such a diagram for the operation of a hypothetical chemical production plant is shown in Fig. 9.2. The diagram plots the probability of an accident (shown here as *cumulative* frequency) against the impact of an event (shown here normalized as the number of resulting fatalities). Cumulative frequency can be understood as the probability that an event occurs with an impact that is equal to or less than a specific level of impact. The impact value along the x-axis could also be represented by, for example, a category such as the number of humans injured, contaminated surface waters (in m^3 or km^2), soil with impaired fertility (in $km^2 \times$ years), or property damage in millions of US dollars, etc.

The acceptability line divides the space into an acceptable and an unacceptable region, with a transition area in between. These acceptability boundaries should be defined during the dialogue, although they are of course not always so simple to set given the different patterns of perception and evaluation between groups and individual stakeholders. The steepness of the line itself is a reflection of society's comfort with risk and can be related to the risk aversion factor (α) in Eq. 9.1.

Once the diagram is established, potential events can then be plotted onto it based on their impact and probability of occurrence. Events that are partially in the transition area between acceptable and not acceptable require careful consideration of the trade-offs at stake. If the event is in the unacceptable region, safety measures

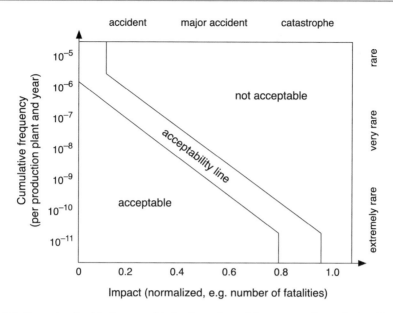

Fig. 9.2 Example of a risk diagram with the dimensions of frequency and magnitude of impact. The diagram shows the set acceptability line defined through societal dialogue that illustrates the acceptable and not acceptable risks for society. The values included in this example are adapted from the Swiss Federal Office for the Environment (2018)

need to be taken to reduce the risk accordingly. If no safety measures can be implemented to make the technology acceptable, an alternative technology needs to be found.

Risk diagrams and matrices are often used to help map the results of risk-dialogue and to support discussions related to legal compliance. Depending on the stakeholders involved in the dialogue, the creation of such a diagram can be representative of the societal acceptance at varying scales (e.g., local, regional, national, or potentially international). Risk diagrams and matrices inherently focus on understanding and mapping the acceptability of existing *risks*; however, they do not support stakeholders in also considering the *benefits* of the technology in question.

9.4   Structure of Societal Dialogue

The structure of a risk-benefit dialogue can vary depending on when and why it is used. It can generally be applied to support three specific types of cases: (1) to help maintain day to day operations (the *normal case*), (2) to help improve relations in the case of a conflict (the *conflict case*), or (3) as a response to a damaging accident (the *accident case*). There are specific aspects to consider in facilitating societal dialogue in each case.

Figure 9.3 depicts a generic structure for facilitating a risk-benefit dialogue. The first step is to initiate the dialogue and set its goal and scope. To do this, relevant stakeholder groups need to be identified, and their values and interests should be clarified. One way to start a dialogue is to organize either *expert workshops* with key stakeholders or *citizen panels* with concerned laypersons in the community. A moderator should be chosen to help facilitate the dialogue that is acceptable for and considered trustworthy by all of the stakeholders involved.

The EU-funded PROSO project (PROSO Consortium, 2019) investigated how to encourage citizens and third-sector organizations such as nongovernmental organizations (NGOs) to participate in research and innovation within society.

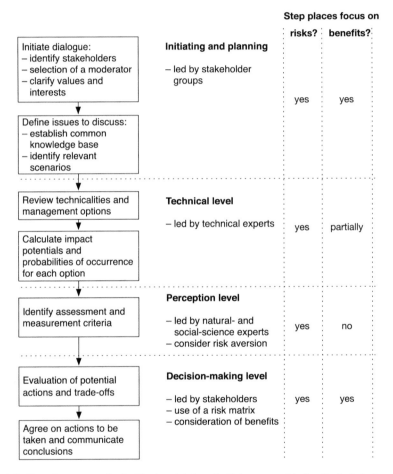

Fig. 9.3 Generic structure for facilitating risk-benefit dialogue considering the technical, perception, and decision-making levels

During the project, several citizen panels were organized in multiple European countries, and a report (PROSO Consortium, 2016) was published detailing the planning process for the meetings, guidelines for communicating with invited citizens, and the meeting objectives. This report can serve as a helpful resource.

In a second step, a common knowledge base needs to be created and shared among the participants to define issues to discuss. Important is to ensure mutual respect for the different types of knowledge available, all of which should be considered. A focus in this step should be placed on searching for and identifying relevant, critical scenarios that pose a risk. Next, technical options for the management of risks should be defined. This should result in a set of options for all stakeholders to later consider, and it is followed by calculating the impact potentials and probabilities of occurrence for each technical option, including calculation of residual risks. Assessment criteria are then developed to support the evaluation of the impact potential and probabilities of occurrence.

Finally, all stakeholders need to consider all of this generated information, discuss potential trade-offs with one another, and arrive at conclusions about the acceptability of a risk and actions to be taken. These conclusions should then be communicated to other stakeholders not present during the dialogue.

As mentioned above, there are generally three cases in which risk-benefit dialogue can be used, and small adjustments might need to be made to the steps shown in Fig. 9.3 accordingly. These are the *normal case*, the *conflict case*, and the *accident case*, and each is further introduced in the following sections.

9.4.1   The Normal Case

In the *normal case* of day-to-day business operations, the generic structure of a risk-benefit dialogue should be used as an ongoing learning and negotiation process. The goals of such a risk-benefit dialogue include:

- to understand society's perception of existing benefits and risks, including the level of objectivity and rationality used
- to exchange factual information and experience about the benefits and risks with stakeholders
- to clarify the protection goals and the conditions for maintaining acceptance within society

New products or changes in a company's operations should initiate a dialogue with stakeholders to ensure that any societal concerns are understood. Doing this successfully also depends on both a company's subject-specific expertise as well as its credibility. Credibility is earned, for example, through demonstrating:

- Consistency between words and actions
- Honesty, openness, and equality
- Patience and comprehensibility
- Acceptance of other points of view
- Rationality and transparency

Standard tools to help the chemical industry be more transparent, trustworthy, and understandable to the public and the media include, e.g., publishing annual environmental and safety reports, reporting accidents and near-accidents as well as the measures taken to address them, communicating deficiencies and how they are corrected, regular media contributions, development of informative websites, organizing neighborhood networks, and offering factory tours. All efforts to initiate dialogue should be proactive, and special attention should be given to also considering and communicating risks from chronic and non-visible exposures.

9.4.2 The Conflict Case

Conflicts can occur following a decrease in societal acceptance of a company, its products, or a technology, and it can lead to public backlashes such as protests, boycotts, and legal action. Given the high stakes that can be at play, communication during such conflicts can be made even more difficult by the emotions that may be involved. While such emotions can complicate the dialogue, it is important that they are actively heard and not seen as irrelevant or used to discredit stakeholders.

The goal of risk-benefit dialogue in this *conflict case* is to understand the terms and conditions under which the benefit of a product or process outweighs the perceived risk for the vast majority of stakeholders.

Here too, a proactive approach following the generic structure of a risk-benefit dialogue is important, and the earlier conflict management begins, the more effective it will be. Conflicts can usually be classified within one of the three levels shown in Table 9.1.

Table 9.1 Different levels of conflict leading to a lack of acceptance of chemical products and processes

Conflict level	Challenges to acceptance
Generic technology	Questioning how desirable it is (e.g., genetic modification)
Specific location	Questioning the societal (spatial) distribution of risks and benefits
Specific product/process	Questioning the safety of the technical design and use/operation

To address concerns regarding a generic technology (such as nuclear power or genetically modified organisms in agriculture), a risk-benefit dialogue can be very costly, and success is likely only possible in the longer term after further discussion and testing. For concerns related to individual products or processes, it is usually much easier to reach an agreement and address stakeholders' concerns.

To gain societal acceptance in the event of a conflict, a participatory approach involving those potentially affected is crucial. For example, dialogue based on the concept of *critical rationality* (Renn, 1993) can help, and success relies on the following:

- Recognition of the need for rational discourse between the conflicting parties
- Openness regarding the outcome of the negotiation
- Equal rights for all parties involved in the dialogue
- Comprehensibility and trace-ability of arguments and outcomes

- Formulation of rational knowledge instead of moral categorization
- Agreement on the procedure during and after the discussion

A risk-benefit dialogue addressing an existing conflict takes time, and discussions could last a year or longer. If acceptance is earned in the end, the industry once again has the responsibility to reach the agreed protection goals and maintain this trust in the long term.

9.4.3 The Accident Case

While they should be adamantly and proactively avoided, accidents can unfortunately never be completely ruled out in the chemical industry. If they do happen, credibility and competence during the management of accidents are very important for the long-term societal acceptance and protection of a company's image. Separate from the actual accident itself, the effectiveness of communication during the management of an accident has a decisive influence on the damage ultimately perceived by society. Decisions often have to be made and communicated under extreme time pressure, and the company has to answer to a great demand for information from the public and the media.

The objective of risk-benefit dialogue in the *accident case* is to maintain credibility through timely, open communication that is responsive to the concerns of those affected.

Requirements for successful crisis management include standardized and timely reporting of events to upper management and clear regulation of relevant responsibilities. For acute accidents such as a fire, explosion, or leakage of toxic or environmentally hazardous substances, rapid and well-founded decisions for the safeguarding of protected goods are needed to minimize damage. According to, for example, the Major Accidents Ordinance in Switzerland (FOEN, 2016): Following an accident, the site owner must immediately secure the location of the event, manage the accident, and inform the appropriate government authority. In addition to eliminating the resulting impacts, the site owner must submit a written report to the government authority within three months (including lessons learned and consequences).

In parallel to managing the damage and documenting the accident itself,[1] communication with the affected population and/or the media must be initiated quickly. For this, a proactive communication concept needs to be prepared in advance and be ready for use in case of an emergency. Dialogue in the event of an accident means that company managers need to immediately assume their responsibilities and already think ahead multiple steps. This involves:

Immediately
- being present in person and taking the perception, concerns, and well-being of the population seriously

---

[1] For example, through taking measurements of pollutants spread in air and water.

- personally apologizing to the affected population regardless of the company's legal liability
- providing information on the facts and the planned course of action openly and in a form understandable to lay persons

Longer-Term
- considering the feelings (possibly fear, anger) of the population during risk management, e.g., openly communicating and discussing how to improve safety measures as a result of the accident
- learning from feedback received from the population
- paying attention to long-term effects of the accident such as damage to reputation and societal acceptance (see Fig. 9.4)

9.5   Opportunities for Improvement

For each of these three cases (normal, conflict, and accident), it is essential for chemical companies to understand the position and concerns of relevant stakeholders. Companies need to be clear about their methods and intent, and they need to use reasonable criteria when weighing and communicating risks and benefits.

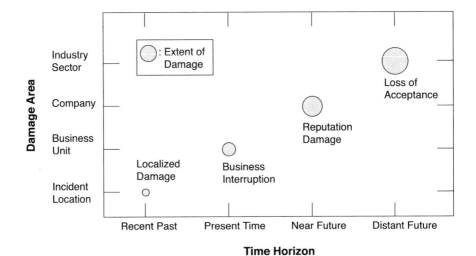

Fig. 9.4 Plot of time horizon vs. potential extent of damage following an accident. Note that in contrast to direct facility damage and business interruptions, damage to reputation and loss of societal acceptance cannot easily be insured. Companies may also be held liable for the immediate or long-term damage to humans or the environment caused by the accident

While the chemical industry has invested in outreach and communication through individual corporate initiatives and sector-wide programs such as Responsible

Care® (ICCA, 2019), there is still an ongoing need to continue developing a deeper dialogue within society. A qualitative review of four chemical companies (Burningham et al., 2007) explored how company executives characterized the public and related this to the company's approach to public engagement and communication. Often, the study found that companies viewed the public as *consumers* of their products or as *neighbors* to their manufacturing sites. In these roles, the chemical industries viewed the public as being important to (1) understand the public's consumption patterns and interest in the industries' products or (2) avoid their industry having negative impacts on the community that could lead to conflicts. However, in this way, the public was not seen as providing the industry with any *knowledge*, and overall the review found the companies to have little interest in engaging the public more widely or earlier in the design process.

The exception to this, the study found, were the few very large corporations that were able to allocate the additional time and financial resources to invest in initiating what are often costly stakeholder dialogues.

The study argues that the chemical industry will need to continue shifting its perception of the public from just consumers and neighbors to instead considering them as valuable stakeholders that through dialogue can provide knowledge and help create benefits. The many smaller companies will need to work toward following the model of their larger competitors and expand their efforts to promote societal dialogue.

Other stakeholders can also improve their contributions to societal dialogue. Hartings and Fahy (2011) argue that chemists (and other scientists for that matter) are themselves important voices in the risk-benefit discussions within chemistry. However, scientists have classically not worked on actively communicating their knowledge and research to nonexperts and instead rely on their scientific publications or on scientific journalists to do this for them. Chemistry, unlike the fields of evolutionary biology and cosmology, also generally lacks inherent narratives that help with storytelling. To help scientists improve communication of their research, R. Olson has written a couple of books on the topic (Olson, 2008, 2015), and Hartings and Fahy recommend scientists consider the following five strategies to help with this:

- *Practice research-driven communication:* Use research from existing communications studies and science communication to understand public stakeholders and how they receive information.
- *Understand the audience:* Consider that there are many types of audiences with different levels of background knowledge and levels of interest in the topic. Aim to understand the audience's attitudes, values, and beliefs about the topic.
- *Participate in the new communication landscape:* Relying on science journalists is a weakening trend, and scientists often have their own social media channels and platforms to communicate and share information. Scientists need to be able to adapt to this modern environment.

- *Tie chemistry to society:* Connect the topic with wider societal issues or broader themes. The audience is more likely to be interested if they can better relate the topic to other issues of interest.
- *Frame key messages to prompt engagement:* Key messages need to be presented in frames (or story lines) that help to simplify complex issues. Consider that different contexts can also be viewed differently (e.g., gene-editing technology may be viewed more positively for applications to improve human health than to modify agricultural products).

Furthermore, the PROSO project identified a set of potential barriers to the participation of third-sector organizations in research and innovation, and it provided specific recommendations for policy makers, research funders, and research organizations to overcome these (PROSO Consortium, 2018). Put into the context of a societal dialogue, the six identified barriers to the participation of citizens and third sector actors include:

- *Lack of relevance:* not perceiving their engagement in a discussion as relevant to their interests, concerns, or goals
- *Lack of impact:* not accepting an invitation to engage if they expect the discussion to have not enough impact
- *Lack of trust and critical views of others:* not engaging when they do not trust the agendas of the organizations running the discussion or of other stakeholders in the discussion
- *Lack of knowledge and skills:* not engaging due to a fear they lack the knowledge and skills needed to participate
- *Lack of time and finances:* not engaging because they do not have the time and/or financial resources necessary to participate
- *Lack of legitimacy:* choosing to not engage because they doubt the legitimacy of the discussion or their involvement in it

By exchanging and rethinking knowledge and different patterns of perception, societal dialogue can help create a shared understanding of a risk and its corresponding benefit. This can lead stakeholders to collectively make responsible decisions about the use of existing and, perhaps more importantly, new chemical technology in society. Working to address each of these six barriers listed above can help to ensure that relevant stakeholders are increasingly part of such discussions.

References

Rise of the citizen scientist (2015) Nature 524(7565):265–265, https://doi.org/10.1038/524265a

Beser A (2016) How Citizen Science Changed the Way Fukushima Radiation is Reported. https://blog.nationalgeographic.org/2016/02/13/how-citizen-science-changed-the-way-fukushima-radiation-is-reported/

Brook RK, McLachlan SM (2008) Trends and prospects for local knowledge in ecological and conservation research and monitoring. Biodiversity and Conservation 17(14):3501–3512, https://doi.org/10.1007/s10531-008-9445-x, http://link.springer.com/10.1007/s10531-008-9445-x

Bürger schaffen Wissen (2019) About Us. https://www.buergerschaffenwissen.de/en

Burningham K, Barnett J, Carr A, Clift R, Wehrmeyer W (2007) Industrial constructions of publics and public knowledge: a qualitative investigation of practice in the UK chemicals industry. Public Understanding of Science 16(1):23–43,https://doi.org/10.1177/0963662506071285, https://doi.org/10.1177%2F0963662506071285

die Debatte (2019) Die Debatte - Was steckt dahinter? https://www.die-debatte.org/

Flint Water Study (2016) FlintWaterStudy.org Guide. http://flintwaterstudy.org/guide-to-flintwaterstudy-org/

FOEN (2016) The Major Accidents Ordinance takes shape. https://www.bafu.admin.ch/bafu/en/home/topics/major-accidents/dossiers/schweizerhalle-chemical-accident/major-accidents-ordinance-tkes-shape.html

Hartings MR, Fahy D (2011) Communicating chemistry for public engagement. Nature Chemistry 3(9):674–677, https://doi.org/10.1038/nchem.1094, http://www.nature.com/articles/nchem.1094

ICCA (2019) Responsible Care. https://www.icca-chem.org/responsible-care/

Müller M, Schneider M, Salathe M, Vayena E (2019) Combining Crowdsourcing and Deep Learning to Assess Public Opinion on CRISPR-Cas9. BioRxiv https://doi.org/10.1101/802454, https://www.biorxiv.org/content/10.1101/802454v3

Olson R (2008) Don't Be Such a Scientist: Talking Substance in an Age of Style. Island Press

Olson R (2015) Houston, We Have a Narrative: Why Science Needs Story. University of Chicago Press

PROSO Consortium (2016) Manual - Citizen Panel Meetings http://www.proso-project.eu/wp-content/uploads/wp4_manual_citizen_panels.pdf

PROSO Consortium (2018) Engaging Society for Responsible Research and Innovation Lowering Barriers - Innovating Policies and Practices p 53, http://www.proso-project.eu/proso-support-tool-2018.pdf

PROSO Consortium (2019) PROSO Project. http://www.proso-project.eu/

Renn O (1993) Risikokommunikation. Christoph Merian Verlag

Stiftung Risiko-Dialog (2019) Risk Dialogue Foundation. https://www.risiko-dialog.ch/

Swiss Federal Office for Public Health (2018) Dialogue-based events. https://www.bag.admin.ch/bag/en/home/gesund-leben/umwelt-und-gesundheit/chemikalien/nanotechnologie/dialogforen-nanotechnologie.html

Swiss Federal Office for the Environment (2018) Beurteilungskriterien zur Störfallverordnung (StFV). Tech. rep., https://www.bafu.admin.ch/bafu/de/home/themen/stoerfallvorsorge/publikationen-studien/publikationen/beurteilungskriterien-zur-stoerfallverordnung-stfv.html

Wissenschaft im Dialog (2019) About Us. https://www.wissenschaft-im-dialog.de/en/about-us/

Wylie S, Shapiro N, Liboiron M (2017a) Making and Doing Politics Through Grassroots Scientific Research on the Energy and Petrochemical Industries. Engaging Science, Technology, and Society 3(0):393, https://doi.org/10.17351/ests2017.134, https://estsjournal.org/index.php/ests/article/view/134

Wylie S, Wilder E, Vera L, Thomas D, McLaughlin M (2017b) Materializing Exposure: Developing an Indexical Method to Visualize Health Hazards Related to Fossil Fuel Extraction. Engaging Science, Technology, and Society 3(0):426, https://doi.org/10.17351/ests2017.123, https://estsjournal.org/index.php/ests/article/view/123

Part III

Implementation

Illustrative Case Study 10

10.1 Introduction

This chapter provides an illustrative case study using the plant protection product line Bion® produced by agrochemical company Syngenta. These products are used to protect against fungal diseases in sunflowers, cereals, vegetables, and specialty crops such as bananas. The active ingredient within Bion® products is acibenzolar-S-methyl (ASM) (CAS Registry Number 135158-54-2).

This case study provides exercises in the form of tasks that apply many of the quantitative and qualitative aspects within life cycle assessment, product risk assessment, process risk assessment, and thermal process safety introduced in the earlier chapters. If worked through in a team and as a structured assignment with a deliverable report and presentation, it aims to help develop multidisciplinary professional teamwork, time management, and presentation skills required to work on similar projects in the field.

Relevant data and references for working through the questions are provided; however additional, supporting data can also be found in external sources. The reader's aim should be to work through each of the tasks with the help of the guiding questions provided. It is important to document each step taken and be able to make and justify any assumptions. A set of solutions are provided in Appendix G. However, it is important to note that some of the tasks and specific questions have no single correct answer and instead ask the reader to qualitatively assess a given aspect. In such cases, the solutions provided give an example answer that covers the key aspects but may not be the only solution. Due to the wide range of potential data that could be used in some exercises, specific data sets have been provided. This has simplified some of the calculations made here, and readers are encouraged to explore the data sources on their own and consider how the use of alternative data could affect the results.

This case study was developed in collaboration with Syngenta Crop Protection AG, Basel, Switzerland. Typical data and methods were chosen to provide illustrative, simplified examples for the risk assessment exercises. The data and information presented do not necessarily reflect the current production process of acibenzolar-S-methyl and Bion® products, and the mass flows across the different process steps are simplified and do not balance completely. Further, the methods used here are not meant to be a comprehensive reflection of the much more in-depth risk assessments conducted and risk management measures taken by companies on manufacturing processes and for the use of crop protection products. The primary aim of this case study is to serve as an exercise that makes use of the tools and methods presented in earlier chapters, and it does not represent a judgment of, or provide any recommendations regarding, Bion® products.

Before starting, it will be helpful to first become familiar with plant protection products in general and to understand the intended use of Bion®. As a first step, perform a search for information related to the use of plant protection products in the European Union via EU government websites (see, e.g., European Commission, 2019b; EFSA, 2019). Then review the material safety data sheets (MSDS) and labeling information specifically for Bion® (see Syngenta, 2018). Once these are in hand, complete this first task:

Task: Summarize the application of and some possible challenges involved in using plant protection products. Consider the following guiding questions:

(a) What are plant protection products? Why are they used?
(b) Which properties in plant protection products are desired and which are undesired? Why?
(c) How does Bion® work, and what advantages does it offer according to the manufacturer?

10.2   Process Risk Assessment

To better understand Bion® as a product, it is helpful to take a closer look at how it is produced. The overall synthesis route for the production of the active ingredient ASM is shown in Fig. 10.1. In this section, the procedure of process risk assessment (as introduced in Chap. 7) will be applied specifically to one reaction step within the synthesis route of ASM: the reaction of 3-chloro-2-isopropylthio aniline with copper(I) cyanide to give 3-amino-2-isopropylthio benzonitrile, as shown in Fig. 10.2.

This is a nucleophilic aromatic substitution reaction called the Rosenmund-von Braun reaction.[1] The reaction mechanism corresponds to a two-stage addition-elimination, where copper cyanide forms a bond to the aryl halide through oxidative addition, and copper chloride then leaves the molecule via reductive elimination.

---

[1] In a Rosenmund-von Braun reaction, an aryl halide reacts to an aryl nitrile in the presence of copper(I) cyanide in a polar high-boiling solvent such as 3-methylpyridine (Zanon et al., 2003).

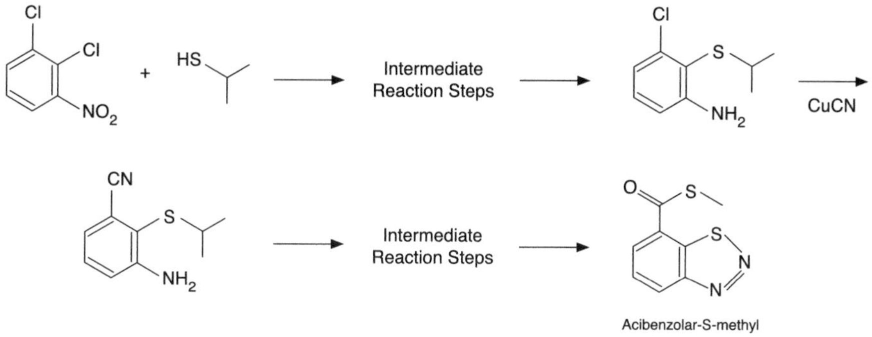

Fig. 10.1 The synthesis route of acibenzolar-S-methyl. Some steps have been redacted, and only steps relevant for the case study are shown

After completion of the reaction, the copper is separated from the reaction mixture by adding sodium sulfide. Sodium sulfide and copper chloride react to give copper sulfide, which precipitates out of solution and is separated by filtration. This is known as the "work-up" and is carried out to purify the product.

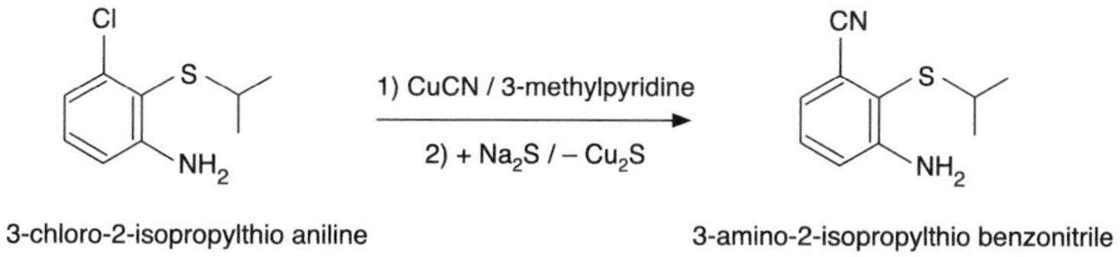

Fig. 10.2 Synthesis reaction (1) and work-up (2) of 3-amino-2-isopropylthio benzonitrile

A known hazard of this process is the formation of hydrogen cyanide (HCN) during the work-up reaction with sodium sulfide. The following subsections go through each of the steps of process risk assessment as introduced in Chap. 7.

10.2.1 Scope, System Definition, and Data Collection (Step 1)

For this first step, assume that the following information has already been collected.

10.2.1.1 Scope

The scope of this risk assessment is to assess and reduce the risk related to the reaction synthesis and work-up of 3-amino-2-isopropylthio benzonitrile during an early stage of process development.

10.2.1.2 System Definition

The system boundaries for this process step are defined by the reactor where the reaction and work-up process are carried out. Energy carriers used are also to be considered, but the storage tanks for the different substances are not part of the system under review. See Fig. 10.3.

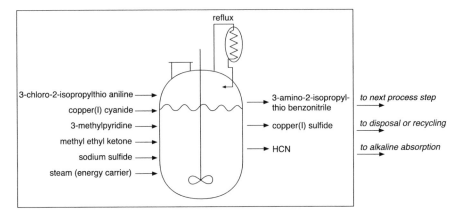

Fig. 10.3 System boundary for the risk assessment of the synthesis reaction and work up of 3-amino-2-isopropylthio benzonitrile

10.2.1.3 Data Collection

Reaction and Process Conditions

The reaction of 303 kg of 3-chloro-2-isopropylthio aniline with 200 kg of copper(I) cyanide, and 210 kg of 3-methylpyridine as a solvent is carried out in batch mode under reflux conditions and stirring at atmospheric pressure for approximately 10 hr.

Although exothermic, the reaction takes place at a high temperature of 190 °C due to a large activation barrier.[2] An additional reason for the high temperature is the autocatalytic nature of the reaction, as there is an induction period that might be positively influenced by higher temperatures (Wang, 2010).

A reflux condition is used for temperature control. Operating at 10 °C below the boiling point of the mixture ensures process safety by avoiding the excessive formation of vapors, which might exceed the capacity of the condenser and then lead to pressure build-up and difficulties to control the temperature. A failure in the temperature control during this exothermic reaction could then lead to a thermal runaway.

---

[2]The process temperature of 190 °C can be achieved at atmospheric pressure (due to the high salt content) even though the boiling points of the product and of the solvent 3-methylpyridine are lower (105 °C and 144 °C, respectively), since the reaction mixture has a boiling point of 200 °C.

After completion of the reaction, the mixture is cooled down and diluted with 180 kg of methyl ethyl ketone. Subsequently, sodium sulfide (Na$_2$S) is added to the reaction mixture in semi-batch mode to react for 2 hr and precipitate the copper(I) as copper sulfide (Cu$_2$S), which can then be separated from the reaction mixture and disposed of or recycled accordingly.

During this exothermic second process step, the heat production rate is controlled by the addition rate of Na$_2$S, which adjusts the reaction rate to the cooling capacity of the reactor. During this process, 2 kg of hydrogen cyanide (HCN) are formed as a side product. Figure 10.4 illustrates the synthesis reaction and precipitation reaction that are within this process step.

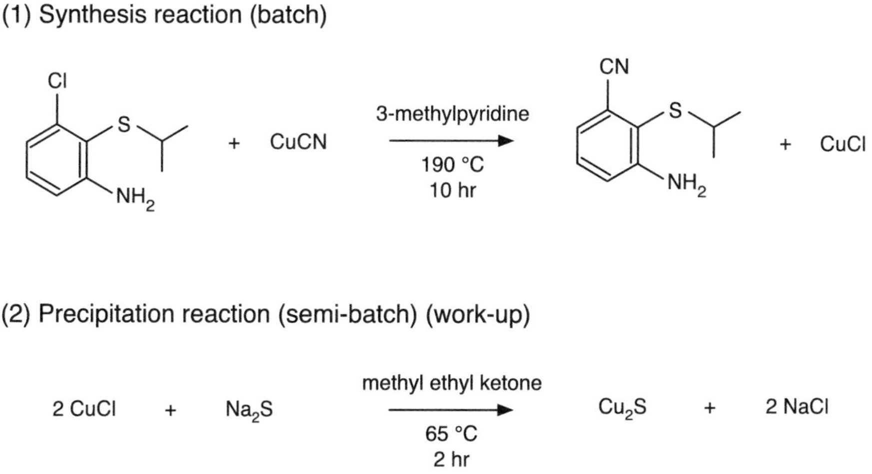

(1) Synthesis reaction (batch)

(2) Precipitation reaction (semi-batch) (work-up)

Fig. 10.4 The synthesis reaction (1) and precipitation reaction (2) taking place in the reactor as they are considered within this risk assessment

Physicochemical Properties and Chemical Hazard Data

Table 10.1 provides physicochemical properties and fire and explosion hazard data for the substances involved in the reaction step. Table 10.2 provides thermal hazard data for the synthesis reaction step and the precipitation reaction step. Table 10.3 provides a reactivity interaction matrix for the synthesis reaction step and the precipitation reaction step. Under acidic conditions, CuCN reacts to form HCN, and Na$_2$S reacts to form H$_2$S. HCN and H$_2$S are poisonous, flammable, and explosive.

Standard Operating Procedure

Written in the form of a standard operating procedure, the steps within the reaction process are:

1. Inertization of the reactor with N$_2$
2. Add 303 kg of 3-chloro-2-isopropylthio aniline

Table 10.1 Relevant property and hazard data for the substances involved in the reaction step

Parameter	3-Chloro-2-isopropylthio aniline	Copper(I) cyanide	3-Methyl-pyridine	3-Amino-2-isopropylthio benzonitrile	Sodium sulfide[a]	Copper(I) sulfide	Methyl ethyl ketone	Hydrogen cyanide
Chemical formula	$C_9H_{12}ClNS$	CuCN	C_6H_7N	$C_{10}H_{12}N_2S$	Na_2S	Cu_2S	C_4H_8O	HCN
Molecular weight [g/mol]	201	89.6	93.1	192	78.0	160	72.1	27.0
Melting point [°C]	<0	473	−18	83–86	950	1100	−18	−13.4
Boiling point [°C][b]	297.9	–	144	105	–	–	79.6	25.7
Density [g/cm³]	1.19	2.92	0.95	–	1.86	5.6	0.95	0.7
Relative vapor density[c]	–	–	–	–	–	–	2.48	0.94
Flash point [°C]	154	–	36	–	–	–	−4	−18
Autoignition temp [°C]	360	–	–	–	–	–	505	538
Decomposition temp [°C]	320	–	–	>340	–	–	–	–

[a]In the reaction, the hydrate form is used ($Na_2S \cdot x\, H_2O$) with a Na_2S content of 60–62%
[b]At 1.013×10^5 Pa (normal atmospheric pressure)
[c]Air = 1

Table 10.2 Thermal hazard data for the synthesis reaction and work-up of 3-amino-2-isopropylthio benzonitrile

	Synthesis reaction	Precipitation reaction
Process temperature [°C]	190	65
Boiling point of reaction mixture [°C]	200	80
Heat of reaction [kJ/kg][a]	−181	−110
Heat capacity [kJ kg^{-1} K^{-1}]	1.9	2

[a]Per kg of reaction mixture

Table 10.3 Reactivity interaction matrix of the chemicals involved in the synthesis reaction and work-up of 3-amino-2-isopropylthio benzonitrile

	H$^+$	CuCN	Na$_2$S	Cu$_2$S
HCN			+	+
3-Chloro-2-isopropylthio aniline		(+)		
CuCN	+			
Na$_2$S	+			
Cu$_2$S	+			
3-Methylpyridine				
Methyl ethyl ketone				
3-Amino-2-isopropylthio benzonitrile				
NaCl				

3. Activate stirrer
4. Add 200 kg of CuCN
5. Heat to 100 °C
6. Add 210 kg of 3-methylpyridine
7. Heat to 190 °C and activate the reflux condenser
8. Stir at 190 °C during 10 hr
9. Cool to 65 °C
10. Add 180 kg of methyl ethyl ketone
11. Add a total of 150 kg of Na$_2$S continuously over the course of 2 hr at 65 °C.

Each batch produces 290 kg of 3-amino-2-isopropylthio benzonitrile. 224 MJ of heat in the form of steam is required to heat the reaction mixture from room temperature (25 °C) to 190 °C.

10.2.2 Definition of Safe Process Conditions (Step 2)

Task: Using the information provided in step 1, define the safe process conditions for the system under review using the questions below as guidance.

(a) Regarding fire and explosion hazards:

- Are safety measures necessary? If yes, which type of measures?
- When should the safety measures be applied?

(b) Regarding side reactions (consider the reactivity interaction matrix in Table 10.3):

- Which hazardous side reactions can occur?
- Which safety measures need to be considered?

(c) Regarding temperature control:

- What is the temperature control strategy?
- Why is the synthesis reaction conducted at $190\,°C$?
- Why is the synthesis reaction not operated at the boiling point of the reaction mixture?

(d) Regarding thermal process safety:

- What is the adiabatic temperature rise and the *MTSR*?
- With these two values in mind, which considerations are important with respect to thermal process safety?
- Classify the potential impacts of each reaction, and also identify the criticality class for each. For the synthesis reaction and the precipitation reaction, assume that $T_{24} > MTSR$ and MTT.

10.2.3 Risk Identification (Step 3)

Considering the data collected in step 1 and the safe process conditions defined in step 2, the process risks can be identified with one of the methods introduced in Chap. 7.

Task: Fill in the missing information in Table 10.4 for the identification of risks during the synthesis and precipitation reactions using a HAZOP method.

10.2.4 Risk Analysis (Step 4)

10.2.4.1 Consequence Analysis

To analyze the consequence of an HCN release scenario, consider the potential effects on a neighboring manufacturing site located $500\,m$ downwind from the reactor.

Table 10.4 Incomplete table showing results of a HAZOP study of the synthesis and work-up reaction of 3-amino-2-isopropylthio benzonitrile. Note that this table is not exhaustive, and other process conditions, deviations, and hazards, etc. could be considered

Item	Process conditions[⊥]	Deviation	Process hazard	Process failure	Loss event	Safeguards	Consequences
Synthesis reaction	Process temperature = 190 °C	High temperature	?	?	Thermal runaway	?	HCN release
Synthesis reaction	Atmospheric pressure	High pressure	?	?	Rupture/crack hole in reactor	?	HCN release
Precipitation reaction	Feed flow (Na$_2$S) approximately 105 kg/hr[a]; Process temperature = 65 °C	High feed flow (semi-batch); high temperature	?	?	Thermal runaway	?	HCN release

[⊥] 3-Chloro-2-isopropylthio aniline reaches the flash point at reaction conditions
[a] Depending on the temperature control loop

Task:

(a) Use the Gaussian model for gas propagation by turbulent diffusion given in Appendix D.2 to calculate the maximum air concentration of HCN expected at the neighboring site. Assume that average wind conditions are 1.5 m/s, stable conditions of $\sigma_y = 18$ m and $\sigma_z = 9$ m prevail, and a leak in the reactor releases 2 kg HCN over 1 hr. Considering the obtained result, estimate the consequence of the release scenario using the consequence categories given in Table 7.8 in Chap. 7. Assume a maximum allowed short-term exposure limit of 3.8 ppm HCN (ECHA, 2019b).
(b) Regarding effects on process operators on-site in the chemical plant, how would you estimate the consequence of a release of HCN using the consequence categories given in Table 7.8 in Chap. 7 and the data provided in step 1?

10.2.4.2 Probability Analysis

As a result of the risk identification step, the process deviations and failures that lead to loss events and to the release of HCN are known. Fault tree analysis can be used to estimate the probability of release of HCN during the synthesis step of 3-amino-2-isopropylthio benzonitrile if probability data for the single events are available (from historical records and reliability analysis). This would result in a quantitative probability estimation. The task below, however, will focus on qualitatively assessing the probability.

Task: Figure 10.5 shows an incomplete fault tree with the release of HCN as the top event.

(a) Insert the missing intermediate events and the corresponding Boolean operators (OR and AND). Note that this tree is rather a simplified model and does not contain an exhaustive list of initiating events.
(b) Comment on the assumption used in the fault tree analysis that all events are independent. Is this realistic?
(c) Qualitatively estimate the probability of release of HCN using the probability categories proposed in Table 7.9 in Chap. 7. Format your solution in a table with probability categories assigned to potential: process failures, process deviations, preventative safeguard failures, loss events, mitigative safeguard failures, and resulting consequences.

10.2.5 Risk Evaluation (Step 5)

Task: Based on the outcomes of the risk analysis step, evaluate the level of risk for people at the neighboring manufacturing site and for operators at the manufacturing

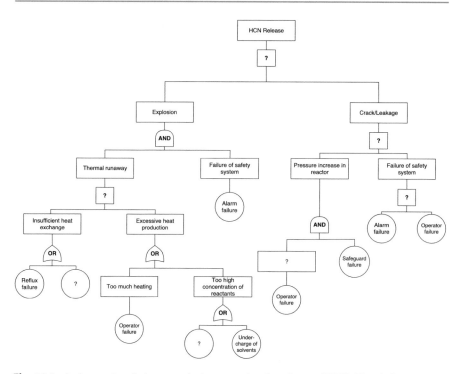

Fig. 10.5 An incomplete fault tree analysis concerning the release of HCN. The circles represent initiating events and the rectangles represent intermediate events

site. To do this, use the risk matrix introduced in Chap. 7 (see Fig. 7.8). Discuss the acceptability of the risk in both cases.

10.2.6 Risk Management (Step 6)

Task: Having evaluated the risks, discuss the following questions:

(a) What safety measures could be taken to reduce the level of risk?
(b) Are there any residual risks that are consciously being accepted?

10.3 Life Cycle Assessment

A life cycle assessment (LCA) can be performed in order to identify the largest impacts of this specific production step of 3-amino-2-isopropylthio benzonitrile (Fig. 10.2). As introduced in Chap. 5, an LCA is composed of four phases. For this case study, the focus is on quantifying and interpreting the impacts of the material flows (production of reactant chemicals and management of resulting waste solvents) and the energy required for heating the reactor.

10.3.1 Goal and Scope Definition (Phase 1)

In the first phase, the goal and scope of the LCA need to be defined. Assume that this LCA was requested to be completed internally within the company in order to understand the source of the largest environmental impacts within this synthesis step and to later compare the total impacts of this step to an alternative design for the step. The impact categories requested to be used are climate change, ecotoxicity and human toxicity, and cumulative energy demand (CED). The system boundaries for the LCA have been requested to ideally cover the production of the reactants, heating of the batch reactor during the synthesis step, and management of wastes via either disposal (incineration) or reuse. For the sake of simplicity, only the waste solvents are quantitatively treated in this case study.

Task: Use what is known about the requested goal and scope of the LCA (from the description above) to answer the following:

(a) What is the objective of the LCA?
(b) What system boundaries can be defined?
(c) What functional unit could be used in the LCA to help compare with future alternative reaction step designs?
(d) Who is the target audience for the results of this LCA? What is their expected level of expertise, and how would this affect how the results of the LCA will be presented?

10.3.2 Inventory Analysis (Phase 2)

The second phase within LCA is to compile a life cycle inventory (LCI) of the system under review. The LCI is a list of the quantified resources and emissions that move into and out of the system boundary. Given the set scope of this particular LCA, this would include the reactants, solvents, heating energy, and products and waste solvents generated (see Fig. 10.3).

Task: Using the information and data from the process risk assessment that describe this process step, construct a simple LCI table that includes (1) the amount of each reactant and solvent used in the production of one batch of 3-amino-2-isopropylthio benzonitrile, (2) the amount of the main product and the waste solvents produced during the process (ignore the generation of Cu_2S, NaCl, and HCN as well as any excess reactants), and (3) the energy required via steam to heat the reaction mass to 190 °C.

10.3.3 Impact Assessment (Phase 3)

In the next phase, the total amounts of material and energy flows from the LCI can be placed within impact categories to estimate their impacts. For this, different LCI databases can be used that provide characterization factors for each flow and for each impact category. Here, impact characterization factors compiled within the Ecoinvent database version 3.5 will be used (Wernet et al., 2016).

Table 10.5 provides characterization factors from the database for each of the material and energy flows considered. The factors describe the midpoint indicators of climate change using global warming potential (GWP) through the IPCC 2013 100-year method, cumulative energy demand, as well as ecotoxicity and total human toxicity using the USEtox model (Rosenbaum et al., 2008; Hauschild et al., 2008; Rosenbaum et al., 2011; Henderson et al., 2011). Notice that the names of the data sets applied for some reagents and solvents do not match the flows within the system being reviewed. This is because the development of such inventory databases is ongoing, and not all products and processes have data available to describe them. In such cases, it can be necessary to approximate the characterization factors by using other (similar) products or processes in the database. In this case, the processes applied may be a significant simplification, and their characterization factors may underestimate the actual impacts of the more complex reactants used in the reaction. Note that the incineration of solvents produces some heat and electricity, which results in them offsetting the need to produce heat and electricity from other energy sources. This credit is considered within the calculation of the characterization factors and results in them being negative.

Task: Using the characterization factors provided in Table 10.5 and the data compiled in the life cycle inventory, calculate the expected impacts. Do this for each material and energy flow separately, and then total the values for each stream type (reactant, waste produced, or energy flow) and for each midpoint indicator.

10.3.4 Interpretation (Phase 4)

After completion of the inventory analysis and impact assessment, the results from the LCA need to be interpreted and used to derive conclusions and recommendations.

Task: Answer the following:

(a) What are the total impacts (climate change, CED, ecotoxicity, and human toxicity) of this process step to produce 3-amino-2-isopropylthio benzonitrile if all of the wastes are assumed to be incinerated? Which material or energy flows have the largest impact? Is this the case across all of the three midpoint indicators?

Table 10.5 LCA impact characterization factors for the mass and energy flows within the life cycle inventory for: climate change (GWP), cumulative energy demand (CED), ecotoxicity, and human toxicity. The factors were extracted from version 3.5 of the Ecoinvent database using the cut off by classification system model and rounded to three significant figures. The activity name of the data set applied to represent each flow is shown as well as its geographical basis, data source, and functional unit. Note that due to limitations in available data, some characterization factors describing similar but not identical activities have to be applied to represent the material or energy flows

Material or energy flow	Activity name of Ecoinvent data set applied	IPCC 2013 GWP 100 yr [kg CO_2-eq]	Total CED [MJ-eq]	USEtox ecotoxicity [CTU_e][a]	USEtox total human toxicity [CTU_h][b]
3-Chloro-2-isopropylthio aniline	"Aniline production" in Europe by Hischier (2007) [per kg]	4.82	104	18.8	1.66×10^{-6}
Copper(I) cyanide (CuCN)	"Copper carbonate production" in Europe by Althaus et al. (2007) [per kg]	2.97	50.2	1170	9.53×10^{-5}
3-Methylpyridine (clean)	"Pyridine-compound production" in Europe by Nemecek (2007) [per kg]	9.21	186	166	4.42×10^{-6}
Methyl ethyl ketone (clean)	"methyl ethyl ketone production" in Europe by Althaus et al. (2007) [per kg]	1.81	62.1	3.60	3.10×10^{-7}
Sodium sulfide (Na₂S)	"Sodium sulfide production" Global, extrapolated by Ecoinvent [per kg]	3.05	42.3	16.9	1.72×10^{-6}
3-Methylpyridine (waste)	"Treatment of spent solvent mixture, hazardous waste incineration" in Europe without Switzerland, extrapolated by Ecoinvent [per kg]	−2.03	−2.95	−0.871	-8.11×10^{-8}
Methyl ethyl ketone (waste)	"Treatment of spent solvent mixture, hazardous waste incineration" in Europe without Switzerland, extrapolated by Ecoinvent [per kg]	−2.03	−2.95	−0.871	-8.11×10^{-8}
Steam	"Steam production, as energy carrier, in chemical industry" in Europe by Althaus et al. (2007) [per MJ]	0.103	1.56	7.73×10^{-2}	8.26×10^{-9}

[a] CTU_e = comparative toxic units for ecotoxicity
[b] CTU_h = comparative toxic units for human toxicity

(b) Assume that, instead of being incinerated, the wastes were recycled and reused in this or other manufacturing processes. Although the relevant characterization factors are not provided in order to make the calculations, comment on how recycling these wastes could potentially affect the LCA's results.
(c) Evaluate the completeness and consistency of the LCA carried out here. What are the limitations and uncertainties of the assessment?
(d) Are there any conclusions or recommendations that could be made based on these results?

10.4 Product Risk Assessment

Following the five steps of product risk assessment introduced in Chap. 6, in this section published data describing the active ingredient acibenzolar-S-methyl (ASM) are reviewed to identify risks the substance may pose to humans and the environment from normal product use of Bion®. Hazards will first be identified by reviewing available product information, and exercises will then be introduced that calculate the estimated environmental exposure to ASM in each compartment as well as both human occupational and dietary consumer exposure to ASM. This will be followed by an effect assessment to identify acceptable levels of exposure, which will then be used together with the estimated exposures to characterize the risks in terms of risk quotients. In the last step, options to manage any nonacceptable risks will be identified.

10.4.1 Hazard Identification (Step 1)

The first step of product risk assessment is the identification of hazards (see Sect. 6.3).

Task: Review and compare the following sources of information on ASM: (1) the conclusion on the peer review of the pesticide risk assessment published by EFSA (EFSA, 2014a), (2) the REACH registration dossier for ASM (ECHA, 2019c), (3) the brief profile for ASM published by ECHA (ECHA, 2019a), and (4) the product labels for Bion® published by Syngenta (Syngenta, 2018). Using these data sources, find out if there are any hazard classifications from the Globally Harmonized System of Classification and Labelling (GHS) assigned to ASM. If yes, how are the hazards defined and described?

Keep these resources easily available as they will be needed for tasks in the upcoming risk assessment steps.

10.4.2 Exposure Assessment (Step 2)

The second step of the assessment is to identify the level of exposure to ASM for the environment and for humans (see Sect. 6.4). In this case study, the focus is placed on (1) using an environmental fate model to predict the environmental exposure and (2) using EFSA calculation tools to estimate both occupational and dietary consumer exposure.

10.4.2.1 Environmental Exposure

In order to evaluate the fate of a chemical in the environment and to predict the resulting environmental concentrations, environmental fate models can be used. In these models, the environment is represented as a collection of well-mixed compartments (i.e., water, soil, and air), and the chemical concentration in each compartment is calculated using partition coefficients and rate constants for transport and transformation processes. Models can vary from level I to level IV, which indicates the environmental processes included and the state of the system considered in the model (see Sect. 6.4.2.2).

For the analysis of ASM, both level I and level III models available within the small-world and small-region modeling tools will be used to investigate the predicted environmental concentration (PEC) of ASM in the soil of an agricultural field and in nearby water bodies. Both models are spreadsheet-based. The model file and an introduction to using it are available online (Scheringer and MacLeod, 2021). Table 10.6 contains information on the properties of ASM relevant for modeling its environmental fate.

Table 10.6 Degradation half-lives and physicochemical properties of ASM

Category	Parameter	Value
Degradation half-lives	$t_{1/2}$ aerobic soil metabolism [hr]	ca. 20[a]
	$t_{1/2}$ degradation in water [hr]	19.4[a,b]
	$t_{1/2}$ atmospheric degradation [hr]	39[a,c]
Physicochemical properties	log K_{ow}	3.1[d]
	log K_{aw}	−5.3[e]
	Molecular weight [g/mol]	210

[a]Values reported by the Australian Pesticides and Veterinary Medicines Authority (2007)
[b]Hydrolysis at 20 °C
[c]Degradation by OH radicals
[d]Value reported by ECHA (2019c) at 25 °C
[e]Calculated from Henry's law constant reported by EFSA (2014a) at 25 °C

Level I Model

A level I model uses partition coefficients to estimate a chemical's environmental distribution under equilibrium. This provides insight into the chemical's distribution in the environment due to thermodynamic drivers only and does not consider flows of the chemical out of the system (e.g., degradation, etc.).

Task: Answer the following:

(a) Use symbolic notation to represent in three equations the concentration of a chemical in water (w), soil (s), and air (a) (e.g., $c_s = \ldots$) in a level I model using only the following symbols:

- Partition coefficients: K_{sw}, K_{aw}
- Volumes of compartments: V_a, V_w, and V_s
- Assume that only the mass of the chemical in soil (m_s) is known.

(b) Use the small-world level I model (Scheringer and MacLeod, 2021) to determine the distribution of ASM into the air, water, soil compartments in Europe. To do this, the following parameters need to be updated within the model:

- Insert the log K_{ow} and log K_{aw} values from Table 10.6.
- Adjust the modeled system to the size of Europe: review external information to set the variables for (1) total surface area [m^2], (2) fraction of area covered by water, and (3) fraction of area covered by soil. If you cannot find these values, default values that can be used are provided in the solutions. Examine the resulting pie chart. According to the level I model, to which compartments does ASM partition?

Level III Model

The environmental concentrations of ASM can also be estimated with a level III model. This model type provides a more realistic estimation of the environmental fate of a chemical by also including the dynamic exchange of materials between compartments. In this model, the system is assumed to be in a steady-state condition where continuous chemical input (emissions) is balanced by its output (via degradation and advection) and all concentrations are therefore time-independent. Similar to the level I model, the system here is composed of agricultural soil, a water body, and the air compartment.

Task: Answer the following:

(a) Write down the first-order differential equations for the change in the mass of a chemical with time ($\frac{dm_{soil}}{dt} = \ldots$) in the three compartments soil (s), water (w), and air (a). When set to zero, these equations represent a level III model. Include only the following processes and symbols (start by identifying the gain and loss processes in each compartment):

- Mass in the different compartments: m_s, m_w, m_a (in mol or kg)
- First-order degradation rate constants in the compartments: k_s, k_w, k_a (in s^{-1})
- Emissions into the compartments: E_s, E_w, E_a (in mol/hr or kg/hr)
- First-order transfer rate constants between the compartments: t_{as}, t_{sa}, t_{aw}, t_{wa}, t_{ws}, t_{sw} (in s^{-1})

(b) In order to run the level III model, the emission rate of ASM is needed in addition to the earlier parameters used for the level I model and provided in Table 10.6. Use the assumed ASM concentration, application rate, and application interval in Table 10.7 to calculate the estimated continuous emission rate [mol/hr] of ASM into 1 hectare of soil during the growing season.

(c) In the small-region model (Scheringer and MacLeod, 2021), update the partition coefficients, degradation half-lives, and fractions covered by water and soil to match those used previously in the small-world model. Update the total surface area parameter to represent 1 hectare of land. Keep all of the other parameters as their default values. Now, obtain the expected distributions in the different environmental compartments using the level III solution. Perform this analysis for the following four emission scenarios using the calculated continuous emission rate:

- Emission only to air (e.g., represents spraying the product from above onto the agricultural field)
- Emission only to water (e.g., represents when the product is accidentally spilled into surface water)
- Emission only to soil
- Emission only to sediments

(d) Comment on the resulting predicted distribution for each of the four emission scenarios. What is the relationship between the emission scenario and the distribution of the chemical in the environment?

(e) What are the assumptions and limitations of the level III model? What is required to obtain a more realistic environmental fate model?

10.4.2.2 Human Exposure: Occupational

Occupational exposure to crop protection products can be assessed using various methods such as modeling (as a lower tier) and field studies measuring real-life operator exposure (as a higher tier). Here, occupational exposure will be investigated using a simple exposure assessment model developed by the European Food Safety Authority (EFSA). This spreadsheet-based tool has been published as a part of EFSA's guidance document on the assessment of exposure to pesticides (EFSA, 2014b). Background information describing the guidance criteria implemented in the tool is provided in the EFSA publication itself, and the tool can be downloaded as a spreadsheet-based file from the publication's supporting information.

Task: Use this EFSA spreadsheet-based exposure assessment tool to complete the following:

(a) Calculate the total acute and longer-term systemic exposures [mg ASM/kg body weight/d] for an operator spraying tomatoes (a fruiting vegetable) with a handheld sprayer. Compare exposure estimates without and with personal protective equipment (PPE). Use the data provided in Table 10.7, and also consider the following:

- ASM is a substance of low volatility and the product Bion® is a wettable granule.
- The reference value non-acutely toxic active substances (RVNAS) is equal to the set acceptable operator exposure level (AOEL) as published in EFSA (2014a).
- An average operator's weight is 60 kg.
- When an operator wears PPE, this includes gloves, workwear, and a hood and visor during both mixing and loading as well as application. A water-soluble bag is used.

Note that Bion's® product label requires appropriate clothing and PPE to be worn during application. The scenario without using PPE is entirely hypothetical.

(b) Which pathway is estimated to be the largest contributor to the systemic longer-term exposure to ASM?
(c) For which application route does PPE provide the most benefit? Which PPE is that?

10.4.2.3 Human Exposure: Dietary

In addition to occupational exposure, dietary exposure of consumers can also be investigated. Here, the aim will be to assess the influence that the amount of pesticide residue remaining on a food product (crop) has on the resulting dietary exposure. These residues are regulated through *maximum residue levels (MRLs).* On the basis of dietary data from the EFSA PRIMo 3.1 model (Anastassiadou et al., 2019) and current MRLs from the EU pesticides database (European Commission, 2019a), the influence of MRLs for ASM on the dietary predicted daily intake (PDI) of different age groups will be investigated. Guidance documentation for the EFSA tool has been published (Anastassiadou et al., 2019), and the tool file itself can be downloaded from that document's supporting information.

Task: Use data from the EFSA PRIMo 3.1 model and EU pesticides database to complete the following:

Table 10.7 Illustrative data describing factors related to operator exposure during application of Bion® with a handheld sprayer with a hydraulic nozzle outdoors for a low-level target such as tomatoes

Variable	Value	Unit
Application		
Bion® maximum application rate	75	g/ha
Concentration of active substance (ASM) in product	500	g/kg
Operator's rate of work	4	ha/d
Minimum application volume	350	L/ha
50% Dissipation time (DT_{50})	1	d
Initial dislodgeable foliar residue	3	$\mu g/cm^2$
Absorption		
Dermal absorption (of product and in-use dilution)	10	%
Inhalation absorption	100	%
Oral absorption	0	%
Scenario		
Application	Outdoor	
Application method	Downward spraying	
Number of applications per season	5	
Buffer strip	2–3	m
Time interval between applications	24	d
Season	Not relevant	

(a) Use Table 10.8 as a template to collect the dietary intake values [g kg_{bw}^{-1} d^{-1}] of the included foods and age groups for both Italy (IT) and the United Kingdom (UK). The data for chronic food consumption from the EFSA PRIMo model should be used (see the "chronic_consumption" sheet of the tool). There may not be data available for all age groups of both countries. In that case, simply extract data for the age groups that are available.

Table 10.8 Incomplete table containing dietary intake values [g kg_{bw}^{-1} d^{-1}] of various foods for the three age groups of infant, toddler, and adult

	Infant	Toddler	Adult
Apples	....	....	....
Apricots	....	....	....
Bananas	....	....	....
Hazelnuts	....	....	....
Lettuce (leaf vegetables)	....	....	....
Mangoes	....	....	....
Peaches	....	....	....
Pears	....	....	....
Spinach and similar	....	....	....
Tomatoes	....	....	....

(b) Find the current maximum residue levels (MRL) for ASM in the EU (use the EU pesticides database European Commission, 2019a) for each of the foods included in Table 10.8.

(c) Carry out a simple calculation to define the predicted daily intake (PDI) [mg kg_{bw}^{-1} d^{-1}] of ASM for each of the three dietary age groups from each of the foods in Table 10.8 for both the United Kingdom and Italy. Use the extracted dietary intake values, and assume the MRL for each food is applied.

(d) Does the daily intake of ASM differ significantly between the various foods? In each country, which of the three dietary groups is predicted to consume the largest amount of ASM through this particular set of foods?

10.4.3 Effect Assessment (Step 3)

In this third step within product risk assessment, the relationship between the level of exposure to a substance and the level of resulting effect is determined (the dose-response relationship; see Sect. 6.5). For ASM, results from published toxicity studies exist that can be used to help complete the effect assessment.

Many of these have been compiled within the EFSA conclusion on the peer review of the pesticide risk assessment of ASM (EFSA, 2014a). In the tasks below, values will be extracted from this document to represent acceptable exposure levels for environmental health using toxicity studies on algae, daphnia, and fish species as well as for human health considering data describing occupational and dietary exposure.

10.4.3.1 Environmental Toxicity
A safe level of environmental exposure can be defined through the use of, e.g., a no observed effect concentration (NOEC) based on ecotoxicity studies and an appropriate assessment factor to define the predicted no effect concentration (PNEC).

Task: Review the published ecotoxicty data in the EFSA conclusion, and extract the following toxicity values for these three aquatic organisms:

- EC_{50} on biomass (EbC_{50}) for the algae species *Scenedesmus subspicatus* from a 72-hr static test
- NOEC for the daphnia species *Daphnia magna* from a 22-day semi-static test
- NOEC for the fish species *Oncorhynchus mykiss* from a 87-day flow-through test

10.4.3.2 Human Toxicity: Occupational
A safe level of occupational exposure during application of Bion® can be defined using an acceptable operator exposure level (AOEL) to ASM.

Task: What is the published AOEL for ASM within the EFSA conclusion?

10.4.3.3 Human Toxicity: Dietary

An acceptable daily intake (ADI) is a measure of the amount of a specific substance in food or drinking water that can be orally ingested by everyone on a daily basis over a lifetime without an appreciable health risk. This can be defined, e.g., for a food additive or residue of a veterinary drug or pesticide.

Task: What is the published ADI for ASM within the EFSA conclusion?

10.4.4 Risk Characterization and Classification (Step 4)

With both some of the potential exposures and reported effects of ASM analyzed, these can now be compared in order to characterize and classify risks that could exist (see Sect. 6.6).

10.4.4.1 Risk to the Environment

Task: Consider the predicted environmental concentration (PEC) in the water compartment as estimated from the small-region model assuming a worst-case scenario that the maximum application rate of Bion® is used, and it is all released directly into the water compartment. Also, consider the EC_{50} and NOEC values of the three aquatic species identified during the effect assessment in the previous step. Using an assumed assessment factor of 10 to consider the most sensitive organism within each species, calculate the PNEC and the risk quotient for this scenario. Is there a risk estimated to be posed in this case according to these toxicity studies at this environmental concentration?

10.4.4.2 Risk to Humans: Occupational

Task: Consider the calculated longer-term operational exposures to ASM from the EFSA tool and the published value of the AOEL identified during the effect assessment in the previous step. For the two operational exposure scenarios calculated using the tool (i.e., with and without PPE), what is the longer-term systemic exposure in % of the AOEL? Is there a risk posed based on these values?

10.4.4.3 Risk to Humans: Dietary

Task: Compare the calculated PDI values of ASM across the different age groups and countries to the published EFSA value of the ADI. Is there a risk posed to any group based on these values?

10.4.5 Risk Management (Step 5)

The final step of product risk assessment is to identify and implement actions or measures to reduce any risks that have been classified as not acceptable (see Sect. 6.7).

Task: Answer the following for the three areas investigated in this case study (risk to the environment, occupational risk, and dietary risk):

(a) Which risks identified during the assessment could be deemed as not acceptable?
(b) What measures or actions could potentially be taken specifically to reduce these risks?
(c) What limitations or uncertainties exist in the risk assessment carried out in this case study?

References

Althaus HJ, Chudacoff M, Hischier R, Jungbluth N, Osses M, Primas A (2007) Life Cycle Inventories of Chemicals. Tech. rep., Swiss Centre for Life Cycle Inventories, Dübendorf

Anastassiadou M, Brancato A, Carrasco Cabrera L, Ferreira L, Greco L, Jarrah S, Kazocina A, Leuschner R, Magrans JO, Miron I, Pedersen R, Raczyk M, Reich H, Ruocco S, Sacchi A, Santos M, Stanek A, Tarazona J, Theobald A, Verani A (2019) Pesticide Residue Intake Model-EFSA PRIMo revision 3.1. EFSA Supporting Publications 16(3):1–15, https://doi.org/10.2903/sp.efsa.2019.EN-1605, http://doi.wiley.com/10.2903/sp.efsa.2019.EN-1605

Australian Pesticides and Veterinary Medicines Authority (2007) Evaluation of the new active acibenzolar-s-methyl in the product Bion plant activator seed treatment. Tech. rep., https://apvma.gov.au/node/13576

ECHA (2019a) Brief Profile: S-methyl benzo(1.2.3)thiadiazole-7-carbothioate. https://echa.europa.eu/brief-profile/-/briefprofile/100.101.876

ECHA (2019b) Hydrogen cyanide: Guidance of Safe Use. https://echa.europa.eu/registration-dossier/-/registered-dossier/14996/9

ECHA (2019c) Registration Dossier: S-methyl benzo(1.2.3)thiadiazole-7-carbothioate. https://echa.europa.eu/registration-dossier/-/registered-dossier/4209/1

EFSA (2014a) Conclusion on the peer review of the pesticide risk assessment of the active substance acibenzolar-S-methyl. EFSA Journal 12(8):3691, https://doi.org/10.2903/j.efsa.2014.3691, http://doi.wiley.com/10.2903/j.efsa.2014.3691

EFSA (2014b) Guidance on the assessment of exposure of operators, workers, residents and bystanders in risk assessment for plant protection products. EFSA Journal 12(10):3874, https://doi.org/10.2903/j.efsa.2014.3874, http://doi.wiley.com/10.2903/j.efsa.2014.3874

EFSA (2019) Pesticides. https://www.efsa.europa.eu/en/topics/topic/pesticides

European Commission (2019a) EU Pesticides Database. http://ec.europa.eu/food/plant/pesticides/eu-pesticides-database/public/?event=homepage&language=EN

European Commission (2019b) Pesticides. https://ec.europa.eu/food/plant/pesticides_en

Hauschild MZ, Huijbregts M, Jolliet O, Macleod M, Margni M, van de Meent D, Rosenbaum RK, McKone TE (2008) Building a Model Based on Scientific Consensus for Life Cycle Impact Assessment of Chemicals: The Search for Harmony and Parsimony. Environmental Science & Technology 42(19):7032–7037, https://doi.org/10.1021/es703145t, https://pubs.acs.org/doi/10.1021/es703145t

Henderson AD, Hauschild MZ, van de Meent D, Huijbregts MAJ, Larsen HF, Margni M, McKone TE, Payet J, Rosenbaum RK, Jolliet O (2011) USEtox fate and ecotoxicity factors for comparative assessment of toxic emissions in life cycle analysis: sensitivity to key chemical properties. The International Journal of Life Cycle Assessment 16(8):701–709, https://doi.org/10.1007/s11367-011-0294-6, http://link.springer.com/10.1007/s11367-011-0294-6

Hischier R (2007) Life Cycle Inventories of Packaging and Graphical Paper. Tech. rep., Swiss Centre for Life Cycle Inventories, Dübendorf

Nemecek T (2007) Life Cycle Inventories of Agricultural Production Systems. Tech. rep., Swiss Centre for Life Cycle Inventories, Dübendorf and Zurich

Rosenbaum RK, Bachmann TM, Gold LS, Huijbregts MAJ, Jolliet O, Juraske R, Koehler A, Larsen HF, MacLeod M, Margni M, McKone TE, Payet J, Schuhmacher M, van de Meent D, Hauschild MZ (2008) USEtox–the UNEP-SETAC toxicity model: recommended characterisation factors for human toxicity and freshwater ecotoxicity in life cycle impact assessment. The International Journal of Life Cycle Assessment 13(7):532–546, https://doi.org/10.1007/s11367-008-0038-4, http://link.springer.com/10.1007/s11367-008-0038-4

Rosenbaum RK, Huijbregts MAJ, Henderson AD, Margni M, McKone TE, van de Meent D, Hauschild MZ, Shaked S, Li DS, Gold LS, Jolliet O (2011) USEtox human exposure and toxicity factors for comparative assessment of toxic emissions in life cycle analysis: sensitivity to key chemical properties. The International Journal of Life Cycle Assessment 16(8):710–727, https://doi.org/10.1007/s11367-011-0316-4, http://link.springer.com/10.1007/s11367-011-0316-4

Scheringer M, MacLeod M (2021) The Small World and Small Region Models – Multimedia Environmental Fate Models for Application in Teaching. https://zenodo.org/record/4438314

Syngenta (2018) Bion 500 FS Seed Treatment. http://www.syngenta-us.com/labels/bion-500-fs

Wang Z (2010) Comprehensive Organic Name Reactions and Reagents. John Wiley & Sons, Inc., Hoboken, NJ, USA, https://doi.org/10.1002/9780470638859, http://doi.wiley.com/10.1002/9780470638859

Wernet G, Bauer C, Steubing B, Reinhard J, Moreno-Ruiz E, Weidema B (2016) The ecoinvent database version 3 (part I): overview and methodology. The International Journal of Life Cycle Assessment 21(9):1218–1230, https://doi.org/10.1007/s11367-016-1087-8, https://doi.org/10.1007/s11367-016-1087-8

Zanon J, Klapars A, Buchwald SL (2003) Copper-Catalyzed Domino Halide Exchange-Cyanation of Aryl Bromides. Journal of the American Chemical Society 125(10):2890–2891, https://doi.org/10.1021/ja0299708, https://pubs.acs.org/doi/10.1021/ja0299708

A Short Chronicle of Chemical Accidents

A

Table A.1 provides an overview of some of the major chemical accidents that have occurred over the past century. Additional information on these and other cases can be found, for example, in the resources provided by the United States Chemical Safety and Hazard Investigation Board (US CSB) (US CSB, 2020b) and in the book *Incidents That Define Process Safety* by Atherton and Gil (2008).

© The Author(s), under exclusive license to Springer Nature Switzerland AG 2021
K. Hungerbühler et al., *Chemical Products and Processes*,
https://doi.org/10.1007/978-3-030-62422-4

Table A.1 Some significant historical accidents in the chemical industry over the past century. Information from Atherton and Gil (2008) and US CSB (2020a)

Date; location; chemicals involved	Accident description	Impact
September 21, 1921; Oppau, Germany; ammonium nitrate	When detonators were used to loosen storage piles of a mixture of 50% NH_4NO_3 and 50% $(NH_4)_2SO_4$ in a silo, the blasting detonated the mixture (4500 t worth). The procedure had been carried out many times before without any incidents with the thought that the mixture was not combustible at this concentration	More than 500 people killed on site and in neighboring villages; entire plant and ca. 700 nearby homes destroyed
June 1, 1974; Flixborough, UK; cyclohexane	In a set of pressurized reactors used in the production of cyclohexanone, one element was replaced by a bypass line. Engineers overlooked limitations in the forces the line was able to handle, and it burst when the reactor was started up again. A flash evaporation of the reactor contents led to a steam cloud explosion	28 employees killed and 36 injured; 53 people in surrounding community injured; entire plant destroyed along with damage to over 1800 nearby homes and 160 businesses
July 10, 1976; Seveso, Italy; chlorinated dibenzodioxin	The site produced the herbicide ingredient 2,4,5-trichlorophenol by partial hydrolysis of tetrachlorobenzene. In the middle of a production batch, the steam to the reactor was switched off, and the workforce went home for the weekend. Approximately 7 hr later, a thermal runaway reaction occurred within the contents left in the reactor, and the rupture disk was activated. This released a mixture of chemicals into the atmosphere including highly toxic 2,3,7,8-tetrachlorodibenzo-para-dioxin. A cloud of white dust containing the chemicals spread over a 17 km^2 area south of the plant and settled onto densely populated villages	More than 200 cases of the skin disease chloracne in nearby residents; 3300 livestock animals died with another ca. 78,000 forced to be slaughtered later; topsoil had to be removed and incinerated; 1 km^2 residential area with more than 700 inhabitants evacuated for months

December 3, 1984; Bhopal, India; methyl isocyanate (MIC)	The plant produced the pesticide Sevin, which involved the highly toxic MIC as an intermediate. Near midnight, water leaked into an MIC storage tank causing an exothermic reaction. At the time, the system's safety layers including a refrigeration unit, air scrubber, and flare were all shut down to help with either cost savings or awaiting maintenance. This resulted in ca. 40 t of MIC to be released into the atmosphere and spread downwind of the plant as a low vapor cloud	ca. 2000 people killed and 100,000 more either blinded or left with other health problems; serious damage to livestock and crops in the area
November 1, 1986; Basel, Switzerland; extinguishing water contaminated with various chemicals	After a fire began in a chemical storage facility in the Schweizerhalle chemical complex, large quantities of extinguishing water used to control the fire were released directly into the Rhine River. The water is estimated to have contained between 13 and 30 t of 90 different chemicals including many pesticides. The wastewater treatment system at the site was significantly under-designed to handle the large volumes of water used to control the fire	Many species of fish and other river organisms were killed up to 400 km downstream of the site; sediment in the river was contaminated down to 14 m depth requiring over 10,000 m^2 of the riverbed to be extracted and incinerated
March 23, 2005; Texas City, US; hydrocarbon mixture from oil refinery operations	An oil refinery owned by BP was undergoing large-scale maintenance across multiple systems. When restarting a hydrocarbon isomerization unit, a distillation tower was flooded with hydrocarbons causing an over-pressurization and release from the vent. The flammable gases and liquids were ignited by a nearby vehicle leading to an explosion. Multiple sensors and alarms were improperly designed or simply failed to alert operators. Cost-saving measures and improper supervisory procedures resulted in improper management of the failures that led to the accident	15 employees and contractors were killed and a further 180 injured; significant damage to the plant

Environmental Distribution and Exposure

B

B.1 Equilibrium Distribution of Substances in the Environment

As introduced in Chap. 6, substances released into the environment will be distributed and transformed through natural processes. The equilibrium distribution of substances between environmental compartments and biota can be described through a set of ratios described in this section. Equations for defining some of the key rate constants that describe the speed of chemical transformation through natural processes are also presented.

B.1.1 Air-Water

$$K_H = \frac{P_{air}}{c_{water}} = R \times T \times \frac{c_{air}}{c_{water}} \tag{B.1}$$

- K_H: Henry coefficient or Henry's law constant, describes equilibrium partitioning between air and water [Pa m^3 mol^{-1}]
- c_{water}: concentration of substance in water [mol/m^3 water]
- c_{air}: chemical concentration in the air [mol/m^3 air]
- P_{air}: partial pressure of substance in air [Pa]
- R: universal gas constant [J mol^{-1} K^{-1}]
- T: temperature [K]

Note: Ionic substances are so strongly hydrophilic that their concentration in the gas phase is negligible.

B.1.2   Water-Solid

$$K_p = K_{oc} \times f_{oc} = \frac{c_{ad}}{c_{water}} \tag{B.2}$$

- K_p: sorption coefficient, describes equilibrium between chemical sorbed to the solid phase and chemical dissolved in water [L water/kg dry solid]
- K_{oc}: partition coefficient between organic solid phase (as carbon) and water; in the case of hydrophobic sorption, K_{oc} [L water/kg organic carbon] can be estimated from the octanol/water partition coefficient, K_{ow} (log K_{oc} = a × log K_{ow} + b, where a and b are coefficients of the regression)
- f_{oc}: fraction of organic carbon in soil solids [kg organic carbon/kg dry solid]
- c_{ad}: concentration of adsorbed substance [mol/kg dry solid]

Note: For ionic substances, Eq. B.2 is not applicable. For the equilibrium distribution of these substances, pH and ionic strength are important influencing factors.

B.1.3   Water-Biota

The concentration of a water-soluble substance in an aquatic organism can be defined using a bioconcentration factor (BCF). As shown in Eq. B.3, the BCF can be calculated (1) as a stationary concentration ratio between biota (e.g., fish) and the surrounding water, as well as (2) a ratio between the respective uptake and elimination rate constants within the organism. The BCF of a chemical is specific for each organism and depends on the organism's lipid content:

$$BCF = \frac{c_{biota}}{c_{water}} = \frac{k_{water}}{k_{elim}} \tag{B.3}$$

- *BCF*: bioconcentration factor [L water/kg biota]
- c_{biota}: concentration of substance in the biota [mol/kg biota]
- c_{water}: concentration of substance in the surrounding water [mol/L]
- k_{water}: pseudo first-order rate constant for uptake of a substance from water [L water/kg biota/d]
- k_{elim}: pseudo first-order rate constant for elimination of a substance from organism [1/d]

Note: Equation B.3 is not valid for ionic and surface-active substances.

B.1.4   Internal Environmental Exposure

In addition to the bioconcentration factor (BCF) defined in Eq. B.3, the internal exposure of an aquatic organism to a chemical can be further described by the concepts of biomagnification and bioaccumulation. As shown in Eq. B.4, the biomagnification factor (BMF) is the ratio of an organism's internal concentration of a chemical to the chemical's concentration in the organism's food source. The BMF represents the ability for the chemical's concentration to increase in organisms up the food chain (at higher trophic levels):

$$BMF = c_{biota}/c_{diet} \qquad (B.4)$$

- *BMF*: biomagnification factor [kg diet/kg biota]
- c_{biota}: concentration of substance in the organism from uptake via the food web [mol/kg biota]
- c_{diet}: concentration in the organism's diet [mol/kg diet]

The bioaccumulation factor (BAF) is defined in Eq. B.5 and is similar to the bioconcentration factor (BCF). The important difference is that the BAF considers all routes of exposure to the chemical (not just from surrounding water):

$$BAF = c_{biota}/c_{surroundings} \qquad (B.5)$$

- *BAF*: bioaccumulation factor [kg surroundings/kg biota]
- c_{biota}: concentration of substance in the organism from uptake via any possible pathway [mol/kg biota]
- $c_{surroundings}$: concentration in the organism's surroundings [mol/kg surroundings]

B.2   Transformations in the Environment

B.2.1   Biodegradation

Biodegradation of chemicals in the environment occurs often in the water, soil, and sediment compartments. This microbial degradation process can be approximated by a pseudo first-order reaction for low environmental concentrations of the chemical undergoing biodegradation:

$$r_{biological} = -\left(\frac{dc}{dt}\right)_{biological} = k_b \times c \qquad (B.6)$$

- r: biodegradation rate [mol m^{-3} s^{-1}]
- c: concentration of chemical undergoing biodegradation [mol/m^3]
- k_b: pseudo first-order rate constant for biodegradation [s^{-1}]

B.2.2   Hydrolysis

Degradation via hydrolysis is strongly pH-dependent (neutral/acidic/basic hydrolysis) and can occur in the water, soil, and sediment compartments. Generally at a constant pH:

$$r_{hydrolysis} = -\left(\frac{dc}{dt}\right)_{hydrolysis} = k_h \times c \qquad (B.7)$$

- r: hydrolysis rate [$mol\,m^{-3}\,s^{-1}$]
- c: concentration of chemical undergoing hydrolysis [mol/m^3]
- k_h: pseudo first-order rate constant for hydrolytic degradation [s^{-1}]

B.2.3   Photolysis

Degradation via photolysis can be caused either directly by interaction of light with the molecules of the substance or indirectly by the interaction of photolytically generated reactive species such as OH radicals (in air) or singlet oxygen (in water). When photolytic degradation in water is considered, the limited penetration depth of sunlight in water needs to be taken into account:

$$r_{photolysis} = -\left(\frac{dc}{dt}\right)_{photolysis} = k_p \times c \qquad (B.8)$$

- r: photolysis rate [$mol\,m^{-3}\,s^{-1}$]
- c: concentration of chemical undergoing photolysis [mol/m^3]
- k_p: pseudo first-order rate constant for photolytic degradation [s^{-1}]

B.2.4   Abiotic Reduction

In the presence of reducing substances, such as Fe or Mn oxides, degradation by abiotic redox reactions can also occur. This can include, for example, reduction of azo compounds, nitroaromatics, halogenated aliphatics, etc.

B.2.5   Additional Comments

Some additional pieces of information to keep in mind regarding degradation processes include the following:

- The total degradation of a substance in the environment across all degradation pathways can often be described by a sum of first-order or pseudo first-order rate constants.
- The degradation rate constant provides no information about the degradation products. While abiotic degradation can often only carry out a single degradation step, biodegradation may lead to complete mineralization of a substance, i.e., the elements C, N, S, and X bound in the chemical degrade into CO_2, NO_3^-, SO_4^{2-}, and X^-.
- Achieving complete degradation often requires a combination of both abiotic and biological stages or the adaptation of microorganisms.

Fires and Explosions

The identification of fire and explosion hazards involves the review of substance properties related to the release of energy and critical gases. These properties can be categorized into either being related to (A) *state characterization* or (B) *dynamics characterization*. Figure C.1 provides an overview of ten key parameters included within these two characterizations.

The state characterization properties include (1) flash point, (2) explosion limits, (3) ignition temperature, and (4) minimum ignition energy. These are depicted within what is known as the triangle of energy release (adapted from ESCIS (1994)). In principle, there is a risk of fire or explosion if fuel, oxygen, and an ignition source come together.[1] State characterization data can, therefore, be used to design conditions that control a substance from being ignited.

Dynamic characterization properties rather describe the hazards posed by a substance when an energy release occurs, and they, therefore, provide relevant information for the design of technical fire and explosion protection systems. These characteristics include (5) fire spreading, (6) explosion dynamics, (7) thermal explosion, (8) deflagration, (9) detonation, and (10) explosive material. Here, the definition of these state and dynamic characterization properties is briefly introduced.

C.1   State Characterization Properties

1. *Flash Point*: the lowest temperature at which a liquid under normal pressure vaporizes to form an ignitable vapor/air mixture. The flash point is used as basis for the GHS classification of flammable liquids, which can be categorized within the following four classes:

---

[1]Note that deflagration could occur even without the presence of oxygen.

© The Author(s), under exclusive license to Springer Nature Switzerland AG 2021   271
K. Hungerbühler et al., *Chemical Products and Processes*,
https://doi.org/10.1007/978-3-030-62422-4

A State characterization properties   **B Dynamics characterization properties**

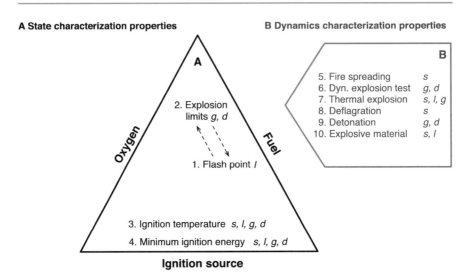

Fig. C.1 Fire and explosion hazard data. The relevant state of the material for each property is shown as: s = solid, l = liquid, g = gaseous, d = dust

- Class I: flash point $<23\,^{\circ}\mathrm{C}$ and boiling point $\leq 35\,^{\circ}\mathrm{C}$
- Class II: flash point $<23\,^{\circ}\mathrm{C}$ and boiling point $>35\,^{\circ}\mathrm{C}$
- Class III: flash point $\geq 23\,^{\circ}\mathrm{C}$ and $\leq 60\,^{\circ}\mathrm{C}$
- Class IV: flash point $>60\,^{\circ}\mathrm{C}$ and $\leq 93\,^{\circ}\mathrm{C}$

Class I substances are extremely flammable, class II are highly flammable, class III are flammable, and class IV are combustible.

2. *Explosion Limits*: the minimum and maximum concentration of a gas or dust in air (e.g., in units of vol % or g/m^3) at which the mixture is explosive. The limits are given for a specific temperature and pressure.

3. *Ignition Temperature*: the lowest temperature of an ignition source at which an optimal fuel/air mixture (solid, liquid, gas, or dust) will ignite.

   A similar property that can be determined is a substance's *autoignition temperature*, which is the lowest temperature at which it will spontaneously ignite in a normal atmosphere without an external source of ignition.

4. *Minimum Ignition Energy*: the amount of energy needed by an ignition source to ignite a liquid, gas, or dust in an optimal explosion mixture (in units of MJ).

C.2   Dynamic Characterization Properties

5. *Fire Spreading*: characterizes how much a solid substance when ignited aids in spreading a fire over an extended area. It can be classified using combustion

numbers, with values of 1–3 representing that it is not fire spreading and values of 4–6 representing that it is.

6. *Dynamic Explosion Parameters*: include the maximum explosion pressure (P_{max}) and the cubic law constant (K_{max}). These are determined through centrally igniting the substance in a closed spherical testing apparatus. The maximum explosion pressure (P_{max}) is the pressure measured for a dust or vapor cloud at its most optimal concentration. According to the ideal gas law, at constant volume Eq. C.1 shows how P_{max} is based on the initial conditions of the pressure (P_0), the amount of combustible material (n_0) in mols, and the temperature of the reaction (T_0). After the explosion, the temperature T_{max} is reached through an adiabatic rise in temperature ($T_{max} = T_0 + T_{ad}$). n_{max} represents the amount of material after the reaction in mols:

$$P_{max} = \frac{P_0 \times n_{max} \times T_{max}}{n_0 \times T_0} \tag{C.1}$$

The cubic law constant (K_{max}) characterizes how fast the pressure from a deflagration or explosion rises in an enclosed vessel. This index is used to properly size explosion vents and design explosion suppression systems. Equation C.2 shows the cubic law for determining the value of K_{max}, where the maximum change in pressure (P) per time is multiplied by the cubic root of the volume (V) of the testing apparatus. Linearly with the scale-up factor ($V^{\frac{1}{3}}$) of the critical equipment, the maximum increase of the explosion pressure is reduced since K_{max} is constant and unique to the mixture of the materials:

$$K_{max} = \left(\frac{dP}{dt}\right)_{max} \times V^{\frac{1}{3}} \tag{C.2}$$

A smaller K_{max} value results in more time until the maximum pressure is reached, and it is important for safety devices to have enough time to properly react before an explosion occurs. As P_{max} is proportional to the normal pressure of the system, critical reactions such as the oxidation of hydrocarbons should ideally take place at the lowest pressure possible. Table C.1 shows an example set of P_{max} and K_{max} for some common substances. The values in this table are based on a P_0 at 1 atm (101,325 Pa). It can be seen that for combustible mixtures at atmospheric pressure, the P_{max} is not extremely high and an appropriately appropriately pressure-resistant device would therefore be feasible to construct (but not for dusts).

The following specific dust hazard classes have also been set to categorize the hazard on the basis of K_{max}:

- ST 0: $K_{max} = 0 \, \text{bar m s}^{-1}$
- ST 1: $1 \leq K_{max} \leq 200 \, \text{bar m s}^{-1}$
- ST 2: $201 \leq K_{max} \leq 300 \, \text{bar m s}^{-1}$
- ST 3: $K_{max} > 300 \, \text{bar m s}^{-1}$

7. *Thermal Explosion*: when the heat generated by a reaction increases its reaction rate (leading to even more heat), a thermal explosion can occur at a determinable ignition temperature when the rate of heat generation exceeds the rate of heat leaving the system (see Chap. 8).

8. *Deflagration*: a locally triggered exothermic decomposition of a solid that can propagate also without oxygen. The reaction propagates at subsonic speeds (0.1 mm/s to 10 m/s) and is driven by the transfer of heat.

9. *Detonation*: a reaction that results in a shock wave that propagates at supersonic speed ($\geq$2000 m/s). The compound-specific maximal detonation rate is a key parameter for explosion control.

10. *Explosive Material*: a substance that causes a sudden release of pressure, gas, and heat when subjected to sudden shock, pressure, or heat. A simple method to screen for chemicals having the formula $C_xH_yO_z$ that may be explosive is through calculating their oxygen balance. Equation C.3 outlines the stoichiometry of the combustion reaction used with the explosive material on the left side of the equation, and Eq. C.4 shows the formula for calculation of the oxygen balance where MW is the molecular weight of the explosive material. The resulting risk of explosion based on the calculated oxygen balance is characterized in Fig. C.2:

$$C_xH_yO_z \rightarrow x\,CO_2 + \frac{y}{2}\,H_2O + \left(\frac{z}{2} - x - \frac{y}{4}\right) O_2 \qquad (C.3)$$

$$\text{Oxygen balance in \%} = 100\frac{16}{MW_{C_xH_yO_z}}\left(z - 2x - \frac{y}{2}\right) \qquad (C.4)$$

Table C.1 Maximum explosion pressure (P_{max}) and cubic law constant (K_{max}) for various substances. For dusts, the ranges indicated are caused by different sizes of the dust particles

Compound	P_{max} [bar]	K_{max} [bar m s^{-1}]
Gases and Vapors		
Hydrogen	7.1	550
Acetylene	10.5	180
Methanol	6.3	46
Dusts		
Aluminum	6.5–13	16–1900
Coal	7.8–8	60–97
Polyethylene	1.3–7.9	4–12

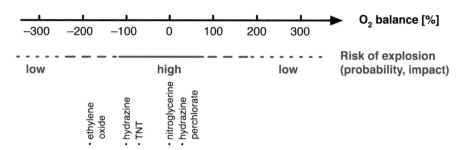

Fig. C.2 Risk of explosion of a substance based on the oxygen balance calculated with Eqs. C.3 and C.4

C.3   Labeling of Fire and Explosion Hazards

The fire and explosion hazards of chemicals are communicated through labeling according to GHS signal words, pictograms, hazard statements and the corresponding hazard (H) codes, and precautionary statements and the corresponding precautionary (P) codes. These can concisely characterize and communicate hazards as well as advise on proper handling, storage, or transportation measures. Table C.2 shows some of the H codes that exist for fire and explosion hazards along with the relevant signal word and pictogram.

Table C.2 Hazard statement (H) codes related to fire and explosion hazards within the GHS classification and labeling system (UN, 2019)

H codes	Meaning	Signal word	Pictogram	Example
H 201	Explosive; mass explosion hazard	Danger		Cellulose nitrate (concentration >12.6%)
H 260	In contact with water releases flammable gases which may ignite spontaneously	Danger		Sodium, potassium
H 271	May cause fire or explosion; strong oxidizer	Danger		Potassium chlorate
H 272	May intensify fire; oxidizer	Danger		Nitric acid (concentration >65%), silver nitrate

C.4   Risk Management of Fire and Explosion Hazards

Following the principle of inherent process safety, the following risk management measures for fire and explosion hazards are suggested in order of decreasing priority:

1. *Prevention or Minimization of Use*

   - Avoid substances with hazardous physicochemical properties (e.g., explosive or combustible substances) and mixtures during manufacturing, transportation, storage, etc.
   - If some are necessary, minimize the quantity used

2. *Compartmentalization*

   - Separate storage areas from other site activities
   - Categorize substances according to hazard classes, e.g., into explosive, oxidizing, flammable according to flash point, fuel consumption, etc.
   - Implement separate storage compartments for explosives, substances in pressure vessels, air-reactive or flammable substances, substances that ignite on contact with water or form flammable gases, oxidizing substances, organic peroxides, and substances that react violently with water or form corrosive/toxic gases, etc.

3. *Inertion*
   Inertion is understood to mean explosion prevention by reducing the oxygen concentration to less than 8% by volume.[2]
   N_2 and CO_2 can be used as inerting gases, but the danger of asphyxiation at less than 17% oxygen by volume needs to be taken into account. Suppression, overpressure, or flow methods can be used, and implementing constant oxygen measuring will ensure that the minimum is maintained. A critical operation is often the substances' insertion into or removal from the inerted area.

4. *Avoiding Ignition Sources*
   Combustible substances exposed to air can ignite due to thermal, mechanical, or electrical energy. Some potential ignition sources are listed in Table C.3 along with corresponding prevention measures. The most important sources of ignition are electrical devices and installed equipment. Electrostatic discharges should also be carefully managed and can occur due to frictional effects (charge

---

[2]Less than 6 vol% for especially critical gases, e.g., H_2 and dusts, e.g., aluminum, paraformaldehyde, 2-naphthol, or biphenyl. Unstable gases such as acetylene and ethylene oxide can decompose exothermically even without oxygen.

separation), e.g., in the stirring, pouring, and spraying of liquids and in the comminution[3] and mixing of solids.

To avoid accumulating electrostatic charges, ensure that all equipment parts are grounded. Areas that are at risk of explosion due to the presence of gases, vapors, or dust should be protected from ignition sources, and ignition by hot surfaces should be avoided through the specification of temperature classes for processing equipment (see Table C.4).

5. *Constructing Explosion Protection*
   This approach attempts to limit the effects of explosions to a manageable level, for example, by using:

   • Explosion-proof construction methods providing building structures with the strength to handle the maximum explosion pressure
   • Explosion pressure relief systems that allow for pressure from an explosion to be released safely
   • Explosion suppression systems that inject a suppressant into the affected equipment when an explosion is detected. The suppressant absorbs the heat from the explosion, thus stopping it.

Table C.3 Examples of ignition sources and measures that can be taken to prevent ignition from these sources

Ignition source	Example	Prevention by
Chemical reaction	Pyrophoric catalyst	Inerting
Fire/spark generation	1. Welding	Organizational safety
	2. Smoking	measures
Hot surfaces	Pipes (steam, heating oil)	Technical protection, e.g., gas-tight insulation
Electrostatic discharge	Flow of liquid or dust	1. Earthing
		2. Inerting
		3. Reducing flow rate (dosing)

---

[3]Comminution is the reduction of solid materials from one average particle size to a smaller average particle size, by crushing, grinding, cutting, vibrating, or other processes.

Table C.4 Temperature classes for processing equipment (such as mills, dryers, etc.) based on the ignition temperature (IEC, 2017)

Ignition temperature of the critical substance [°C]	Maximum allowed surface temperature [°C]	Temperature class
>450	450	T1
300–450	300	T2
200–300	200	T3
135–200	135	T4
100–135	100	T5
85–100	85	T6

Dispersion Models

D

This section presents the basics of two dispersion models relevant for gas propagation from a stationary point source. It also summarizes timescales for global mass transport in the atmosphere and an exchange model for the penetration of gas into a building.

In addition, emissions can also be released from spontaneous point sources and sources elevated far above the ground (e.g., a chimney, thermal chimney effect, etc.). Large obstacles can also stand in the way of an emission and affect its propagation. In these cases, many other and more complex propagation models exist and are introduced in other literature (CPD, 1996; Elvers, 2000).

D.1   Heavy Gas Dispersion

Examples of heavy gas spills include Flixborough in 1974 and Bhopal in 1984 (see Appendix A). In heavy gas dispersion, the spreading of the substance after release is affected by the wind velocity and displacement due to gravity. The validity of this type of model is limited to being within close range of the release site (less than 100 m). The schematic in Fig. D.1 illustrates the model used for heavy gas dispersion.

Conditions for propagation by heavy gas dispersion:

- The density of the gas (ρ_{Gas}) must be greater than the density of the air (ρ_{Air}), meaning that the density difference ($\Delta\rho = \rho_{Gas} - \rho_{Air}$) is the driving force of propagation
- The stationary point source of gas (mass flow $\dot{M}$) is constant and located on a grid at $x, y, z = 0$ (as seen in Fig. D.1)
- x-direction: wind direction (u = wind speed)
- y-direction: propagation occurs due to $\Delta\rho$ at a speed of v_g

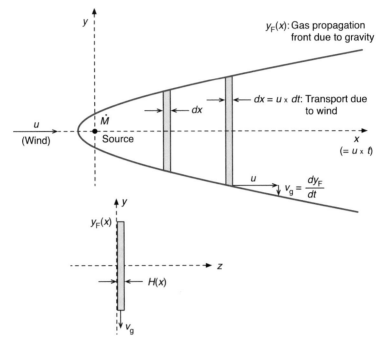

Fig. D.1 Graphical representation of the model for heavy gas propagation. $\dot{M}$ = emission mass flow; u = wind speed; t = time

- z-direction: the thickness of the heavy gas layer in the z-direction ($H(x)$) decreases as the distance from the point source (x) increases. This assumes a stable layering without any air entrainment.
- Ground: plane and fully reflective with no adsorption

Propagation in the y-direction (at a velocity of v_g) can be described by the Bernoulli equation (where g is the acceleration due to gravity):

$$v_g = \left(\frac{2 \times \Delta \rho}{\rho_{\text{Air}}} \times g \times H \right)^{\frac{1}{2}} = \frac{d y_F}{dt} \tag{D.1}$$

At the same time, the law of mass conservation applies to the compartment of width dx with $M = \dot{M} \, dt = 2 \times y_F \times H \times dx \times \rho_{\text{Gas}}$:

$$H(x) = \left(\frac{1}{2 \times y_F} \times \frac{\dot{M}}{\rho_{\text{Gas}}} \times \frac{dt}{dx} \right) = \frac{\dot{M}}{2 \times u \times y_F \times \rho_{\text{Gas}}} \tag{D.2}$$

Applying Eq. D.2 within Eq. D.1 results in:

$$\frac{dy_F}{dt} = \left[\frac{\Delta\rho}{\rho_{Air}} \times g \times \frac{\dot{M}}{u \times y_F \times \rho_{Gas}}\right]^{\frac{1}{2}} \tag{D.3}$$

After integration with $y_F(t = 0) = \pm \delta$ leads to:

$$y_F = \pm \left[\frac{9}{4}\frac{\Delta\rho}{\rho_{Air}} \times g \times \frac{\dot{M}}{u \times \rho_{Gas}}\right]^{\frac{1}{3}} \times t^{\frac{2}{3}} \pm \delta$$

with $x = u \times t$:

$$y_F = \pm\frac{1}{u}\left[\frac{9}{4}\frac{\Delta\rho}{\rho_{Air}} \times g \times \frac{\dot{M}}{\rho_{Gas}}\right]^{\frac{1}{3}} \times x^{\frac{2}{3}} \pm \delta \tag{D.4}$$

where δ is an empirical relationship defined as:

$$\delta \cong 2\frac{\Delta\rho}{\rho_{Air}}\frac{g \times \dot{M}}{\rho_{Gas} \times u^3}$$

In this way, y_F can be expressed as a function of x, and the gas propagation area on the ground can be calculated (assuming there is no air entrainment). Accordingly, the concentration of the gas in the propagation range is constant and equal to the concentration at the source. This assumption is valid only in close vicinity to the source and at low wind speeds (because air turbulence would cause air entrainment at higher speeds).

D.2   Gas Propagation by Turbulent Diffusion

The propagation of flue gas, methane, and many other gases that can be released from a process often occurs through turbulent diffusion. Here, convective propagation is driven by wind speed. The model for this is illustrated in Fig. D.2.

Conditions for propagation by turbulent diffusion according to the model presented here are:

- $\rho_{Gas} \cong \rho_{Air}$ (i.e., no buoyant forces)
- A concentration gradient is the driving force for spreading
- There is a stationary point source of gas where the mass flow rate ($\dot{M}$) is constant and which is located on a grid at $x, y, z = 0$ (as seen in Fig. D.2)
- x-direction: wind direction (u = wind speed, often on average approximately 2 m/s during the day and 1 m/s during the night)
- Ground: planar and fully reflective with no adsorption
- $x > 100$ m and $u > 1$ m/s

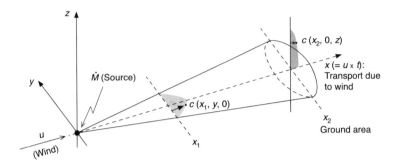

Fig. D.2 Graphical representation of the model for gas propagation via turbulent diffusion. $\dot{M} =$ emission mass flow; $u =$ wind speed; $t =$ time; $c =$ concentration

With diffusion in the x-direction neglected (diffusion $<<$ convection), the law of conservation of mass leads to the assumption that the gas flow through each cross-sectional area x is constant and equal to the flow $\dot{M}$:

$$\dot{M} = \int_0^\infty \left[\int_{-\infty}^\infty c(x, y, z) \times u \, dy \right] dz \tag{D.5}$$

With the assumption that the coefficients for the turbulent diffusion (K_y and K_z) depend only on x, the following is valid at each point:

$$u \frac{\partial c}{\partial x} = K_y \frac{\partial^2 c}{\partial y^2} + K_z \frac{\partial^2 c}{\partial z^2} \tag{D.6}$$

This means that the input by convection in the x-direction is equal to the output by turbulent diffusion (according to Fick's second law) in the y- and z-direction.

The solution of the differential equation (Eq. D.6) with the boundary conditions described is the product of two Gaussian distributions (in the y- and z-direction) where the concentration on the x-axis is proportional to the source strength and inversely proportional to the wind velocity:[1]

$$c(x, y, z) = \frac{\dot{M}}{\pi \times u \times \sigma_y \times \sigma_z} \exp\left\{ -\frac{1}{2} \left[\frac{y^2}{\sigma_y^2} + \frac{z^2}{\sigma_z^2} \right] \right\} \tag{D.7}$$

The distribution variances in the y- and z-directions (σ_y^2 and σ_y^2) can be expressed by using $x = u \times t$ as follows:

$$\sigma_y^2 = \frac{2 K_y(x)}{u} \times x = 2 K_y(x) \times t \tag{D.8}$$

---

[1] Dilution by convection

and:

$$\sigma_z^2 = \frac{2\,K_z(x)}{u} \times x = 2\,K_z(x) \times t \tag{D.9}$$

where σ_y and σ_z are empirical quantities. They depend strongly on the state of atmospheric turbulence, which in turn is mainly determined by the vertical temperature gradient and surface wind (see Table D.1).

Table D.1 Relationship between distributional variance and meteorological conditions

Weather	Atmospheric conditions	σ_y, σ_z
Fog, temperature inversion	Stable (stratification)	σ_z small
Strong solar radiation, strong surface wind	Unstable	σ_y, σ_z large

For considering risk, the main aim is to know the maximum concentration (c_{max}) expected at different distances from the point source (x). This depends on the wind direction. Equation D.7 defines c_{max} (where $x = u \times t$ and $y = z = 0$):

$$c_{max}(x) = \frac{\dot{M}}{\pi \times u \times \sigma_y \times \sigma_z} \tag{D.10}$$

D.3 Timescales for Global Mass Transport in the Atmosphere

See Figs. D.3 and D.4.

D.4 Gas Penetration into Buildings

A simple mass-balance model to estimate the penetration of a gas moving from outside air into a building with volume V can be used:

$$V\frac{dc_i}{dt} = \dot{V} \times c_a - \dot{V} \times c_i \tag{D.11}$$

- V: volume of the building [m^3]
- $\dot{V}$: air exchange rate of the building [m^3/s]
- c_i: pollutant concentration inside the building [g/m^3]
- c_a: pollutant concentration outside the building [g/m^3]

Fig. D.3 Approximate timescales for vertical mixing of gases into different layers of the atmosphere (vertical arrows). The vertical bars show the variability in the height of the boundary layer and the tropopause. Based on estimated values by Jacob (1999)

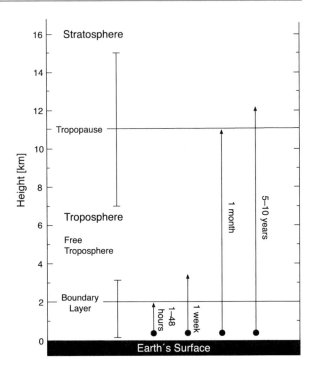

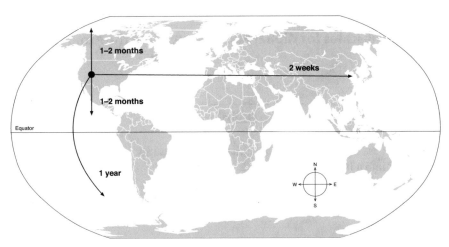

Fig. D.4 Approximate timescales for horizontal transport of gases in the troposphere. Based on estimated values by Jacob (1999)

Integration of this equation with the boundary condition of $c_i(t = 0) = 0$ yields:

$$c_i = c_a \left(1 - e^{-\frac{\dot{V}}{V}t}\right) \tag{D.12}$$

According to this model, a building with low air exchange rate (where $\dot{V}/V$ is ca. $0.5\,\mathrm{hr}^{-1}$) provides significantly more protection than the interior of a car (where $\dot{V}/V$ is ca. $30\,\mathrm{hr}^{-1}$).

Concepts of Probability Analysis

E

E.1 Probability

In the context of process risk assessment introduced in Chap. 7, probabilities are calculated for system failures and are usually expressed in the form of (1) frequency, as the number of events or incidents divided by the exposure period (normally per year) (Eq. E.1); (2) relative frequency, given as a number between 0 and 1, with 0 indicating impossibility and 1 indicating certainty (Eq. E.2); or (3) a conditional probability that an event occurs, given the occurrence of a precursor event (Bayes' theorem) (Eq. E.3):

$$F(A) = \frac{n_A}{t} \tag{E.1}$$

- $F(A)$: frequency of occurrence of event A per unit of time
- n_A: number of outcomes of event A
- t: time period

$$P(A) = \lim_{n \to \infty} \frac{n_A}{n} \tag{E.2}$$

- $P(A)$: probability of event A (relative frequency)
- n_A: number of events A (e.g., number of failures)
- n: number of all events (e.g., total number of operations)

$$P(A|B) = P(B|A) \times P(A)/P(B) \tag{E.3}$$

- $P(A|B)$: the probability of event A given the occurrence of event B
- $P(B|A)$: the probability of event B given the occurrence of event A
- $P(A)$: the probability of event A
- $P(B)$: the probability of event B

© The Author(s), under exclusive license to Springer Nature Switzerland AG 2021
K. Hungerbühler et al., *Chemical Products and Processes*,
https://doi.org/10.1007/978-3-030-62422-4

E.2   Reliability

E.2.1   Equipment Reliability

Reliability is an important aspect of equipment used within chemical processes. A piece of equipment that does not operate as intended poses a significant safety risk. More formally defined, *equipment reliability* is the probability that a piece of process equipment will perform its intended function adequately for a specified period (e.g., time or number of demands) and under specified process conditions.

If n identical equipment units operate without being replaced, after time t, a certain number will have survived ($n_s(t)$), and a certain number will have failed ($n_f(t)$). The probability of survival, or *reliability* $R(t)$, and the probability of failure, $F(t)$ (note that F is used for 'frequency' in Sect. E.1, which is, of course, a different quantity), is defined as:

$$R(t) = \frac{n_s(t)}{n} = 1 - F(t) = 1 - \frac{n_f(t)}{n} \tag{E.4}$$

In order to convert a failure rate into a probability of failure, a continuous probability distribution function can be developed as a function of the instantaneous failure rate (λ):

$$\lambda(t) = \frac{1}{n - n_f(t)} \times \frac{dn_f(t)}{dt} = -\frac{1}{R(t)} \times \frac{dR(t)}{dt} \tag{E.5}$$

This is also equal to $-d\ln(R(t))/dt$. Through integration, this yields the probability of reliability $R(t)$ and probability of failure $F(t)$:

$$R(t) = \exp\left\{-\int_0^t \lambda(t')dt'\right\} \tag{E.6}$$

$$F(t) = 1 - R(t) = 1 - \exp\left\{-\int_0^t \lambda(t')dt'\right\} \tag{E.7}$$

Figure E.1 shows the typical shape of an instantaneous failure rate curve (λ) over time. Equipment reliability can be thought of similarly to the human body's susceptibility to developmental issues and disease. During infancy, the chance for developmental abnormalities or having a weak immune system unable to fight off diseases is high. After development is finished, the probability for health issues to occur remains lower and more constant for some time. Then, as the body approaches old age, the chance for health issues once again begins to increase. The same thinking can be applied to equipment upon its installation and first calibration, through its regular operating life, and then toward the end of its operating life.

Fig. E.1 A typical bathtub
plot of the instantaneous
failure rate (λ) over time. The
integral of $\lambda(t)$ is represented
by the marked area under the
curve

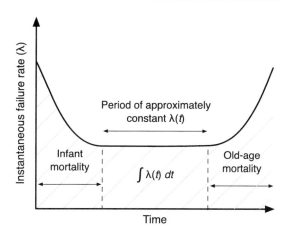

E.2.1.1 Constant Failure Rate Models

Constant failure rate models are used to describe the reliability of equipment that
fails at random intervals and for which the failure rate is constant for long operating
periods. In this case, the reliability at time t ($R(t)$) is the exponentially distributed
probability that a system component does not fail in the time period from 0 to t:

$$R(t) = e^{-\lambda t} \tag{E.8}$$

When $\lambda t \ll 1$, it can be assumed that $R(t) \cong 1 - \lambda t$ (first Taylor approximation).
Logically, the failure probability $F(t)$ is then simply:

$$F(t) = 1 - R(t) = 1 - e^{-\lambda t} \tag{E.9}$$

When $\lambda t \ll 1$, it can be assumed that $F(t) \cong \lambda t$. Both of these functions are
shown in Fig. E.2. The *failure density function* $f(t)$ is then defined as:

$$f(t) = \frac{dF(t)}{dt} = \lambda \times e^{-\lambda t} \tag{E.10}$$

With this density function, the *mean time between failures* (MTBF, $\overline{T}_{BF}$) can be
calculated as:

$$\overline{T}_{BF} = \int_0^\infty t \times f(t)dt = \frac{1}{\lambda} \tag{E.11}$$

E.2.1.2 Reliability of a System with Regular Inspections

Regular inspections are particularly important for safety devices. For process system
components that undergo routine inspections, the probability of failure can be
calculated considering the frequency of inspections ($t_{inspection}$). This *probability of*

failure on demand (PFD) can be defined as:

$$PFD = \frac{t_{\text{unrevealed failure}}}{t_{\text{inspection}}} = \frac{1}{t_{\text{inspection}}} \int_0^{t_{\text{inspection}}} F(t)dt \qquad (E.12)$$

When $\lambda t \ll 1$, this can be simplified to:

$$PFD = \frac{1}{t_{\text{inspection}}} \int_0^{t_{\text{inspection}}} \lambda t \ dt = \frac{1}{2} \times \lambda \times t_{\text{inspection}} \qquad (E.13)$$

This assumes that the time required to repair a failed component is much less than the time that the undiscovered failure has existed ($t_{\text{unrevealed failure}}$). For a low probability of failure on demand which is important for safety devices and ensuring a set safety integrity level (SIL) (see Chap. 7), two parameters can be influenced: (1) a low λ and/or (2) a short interval between inspections.

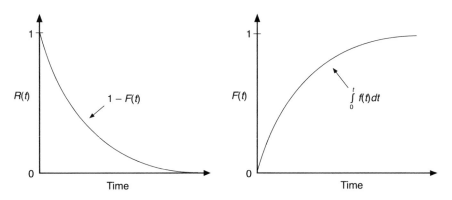

Fig. E.2 Plots of the reliability $R(t)$ (Eq. E.8) and failure probability $F(t)$ (Eq. E.9) functions

E.2.2 Human Reliability

Human error is an important aspect that can contribute to failure within a process and lead to an accident. Human errors are deviations from set operating procedures that can be categorized into failures of either omission, commission, or cognition as shown in Table E.1.

Following the failure of a process step, there is often limited time available for an operator to intervene and prevent (further) damage. The probability of human error decreases following the trend shown in Fig. E.3 when more time is available for intervention.

Fig. E.3 Probability of human failure in responding to a loss event

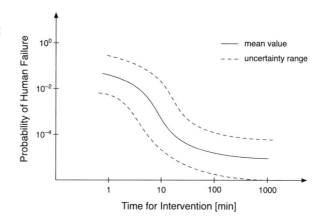

A significant challenge within the field of *human reliability analysis (HRA)* is the limited availability of data (statistics) to describe human reliability in a realistic environment (such as from training in a plant simulator).

Table E.1 Examples of types of human errors

HRA error category	Type of deviation
Error of omission	Omission error: – Task forgotten – Step neglected
Error of commission	Operation error: – Wrong operations – Wrong positions – Wrong command/information – Wrong sequence – Time error (too early/too late) – Quantitative error (too much/too little)
Error of cognition	Perception, diagnostics, decision error: – Wrong correction measure – Mix-up – Ignorance of critical deviation

E.3   Common Cause Failure Analysis

As introduced in Sect. 7.1.2, multiple safeguards are often used in the chemical process industry to reduce process risks. This means that an accident can only occur when multiple safeguards fail. This can result from the independent failure of each safeguard or from a single failure event that causes multiple safeguards simultaneously to fail. The latter is known as a dependent failure event or as a common cause failure.

Risk reduction will be significantly overestimated if common cause failure events are not identified. To illustrate this, consider a protection layer consisting of two temperature switches A and B, which perform the same function (i.e., they are redundant safeguards and using an AND connection). Assuming that the failures of A and B are independent, the probability (P) that both safeguards fail can be calculated using the following expression:

$$P = P_A \times P_B \qquad\qquad (E.14)$$

However, if the failures of A and B are not independent, the probability that both safeguards fail will be higher, as shown by:

$$P > P_A \times P_B \qquad\qquad (E.15)$$

There can be many possible sources of dependency within chemical processes. The CCPS distinguishes the following six types of functional dependency (CCPS, 1999): (1) on common support systems such as a single electricity source, (2) on common hardware (e.g., hardware failures affecting shared equipment), (3) on equipment similarity (e.g., systematic repeated human errors related to a common design, operating procedures, etc. of similar equipment), (4) on a common location (e.g., different equipment in one location can be affected by the same external events, such as environmental conditions, etc.), (5) on a common internal environment for multiple safeguards (e.g., water in an emergency cooling system, air in an instrument air system, etc.), and (6) on common operating/maintenance staff and procedures (e.g., human or procedural errors could occur such as the miscalibration of multiple safeguards).

Some of these dependencies can be addressed through the use of a modified fault tree and event tree analysis known as a *common cause failure analysis*. This is a technique that identifies common cause events and estimates the probability of dependent events using parametric models (CCPS, 1999). In the chemical process industry, the emphasis of common cause failure analysis is usually on systems that rely on multiple layers of protection and high redundancy to achieve process safety. The impact of a common cause failure on the probability of a critical top event is discussed in the next section.

E.4   Probability Analysis for the Grinding Process of Cyanuric Chloride

The example of the wet grinding of cyanuric chloride was introduced in Chap. 7 for the application of process risk assessment. The *loss event* within this process was identified as the decomposition of cyanuric chloride, and this section will show how to estimate (1) its probability of occurrence using a fault tree analysis (FTA) and (2) the probability of the reaction further advancing into a thermal runaway through the use of an event tree analysis (ETA) and considering existing safeguards.

Figure E.4 shows a fault tree with the process deviations that can lead to the decomposition reaction of cyanuric chloride (the *top event*). The process failures leading to the deviation "temperature too high" are also depicted in the fault tree. The events in the tree are linked by OR and AND operators.

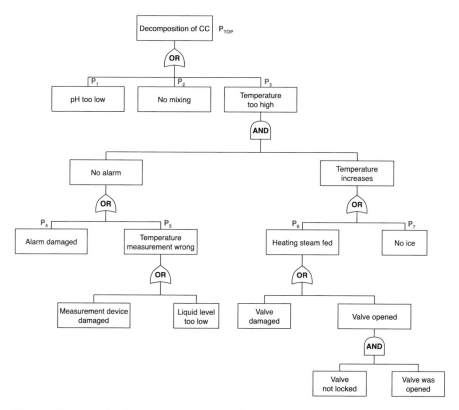

Fig. E.4 Example of a fault tree for the calculation of the likelihood of the decomposition of cyanuric chloride (CC) during the grinding process. The branch for the deviation "temperature too high" is shown. Note that an exhaustive analysis should also include the branches for "pH too low" and "no mixing." The variables representing the probabilities of some events (P_{Top} and P_1 to P_7) are shown

The probability of occurrence of the top event (P_{Top}) is given by the individual probabilities P_1, P_2, and P_3, of the events "pH too low," "no mixing," and "temperature too high." Assuming that these events are independent, using Eq. 7.4, the probability of the top event is given by:

$$P_{Top} = (P_1 + P_2 + P_3) \tag{E.16}$$

The probability of occurrence of the "temperature too high" event (P_3) is given by the individual probabilities P_4, P_5, P_6, and P_7, corresponding to the events "alarm damaged," "temperature measurement wrong," "heating steam fed," and "no ice." This can be calculated using Eq. 7.3 defining AND connections and Eq. 7.4 defining OR connections:

$$P_3 = (P_4 + P_5) \times (P_6 + P_7) \tag{E.17}$$

Important to note within this fault tree is that both the events "alarm damaged" (P_4) and "no ice" (P_7) are dependent on the electricity supply. The probability of having no electricity ($P_{\text{no electricity}}$) can therefore be considered in the calculation of P_3:

$$P_3 = P_{\text{no electricity}} + (P_5 \times P_6) \times (1 - P_{\text{no electricity}}) \tag{E.18}$$

Without electricity ($P_{\text{no electricity}} = 1$), both the safeguards of a temperature alarm and ice cooling system are simultaneously lost. This results in the temperature becoming too high (P_3) and therefore also in the occurrence of the top event (P_{Top}) since P_1, P_2, and P_3 are all OR connected. Without an independent backup electricity supply provided for each of these safeguards, this system is at risk, especially in regions that do not have a reliable electricity supply. Be sure to carefully check the independence of safeguards. They are of no use if they all can fail due to a single underlying issue.

Through the use of Eqs. E.16–E.18, the probability data of all the events can be combined to obtain the probability of occurrence of the top event. As mentioned in Sect. 7.6.2.2, quantitative probability data can be obtained from historical records or reliability analysis.

To consider the impact of mitigative safeguards in this example, an event tree can be used. For illustrative purposes, it is assumed here that the presence of mitigative safeguards does lead to different consequence scenarios.

Figure E.5 shows the event tree starting from the top event (decomposition of cyanuric chloride) and the failures and successes of the mitigative safeguards leading to different consequences ($i = 1$ to $i = 5$).

The probability of the worst-case scenario is given by $P(i = 5)$, which represents the scenario in which all mitigative safeguards fail. Assuming independent events, and using Eq. 7.5, this probability can be expressed as:

$$P(i = 5) = P_{\text{Top}} \times (1 - P_{\text{I}}) \times (1 - P_{\text{II}}) \times (1 - P_{\text{III}}) \times (1 - P_{\text{IV}})$$
$$= P_{\text{Top}} \times \prod_{i=I}^{IV}(1 - P_i) \tag{E.19}$$

To calculate the probability of the worst-case scenario ($i = 5$) in Eq. E.19, the probability of the top event as well as the probability of success of all the existing safeguards (P_{I} to P_{IV}) need to be known. This example yet again highlights the

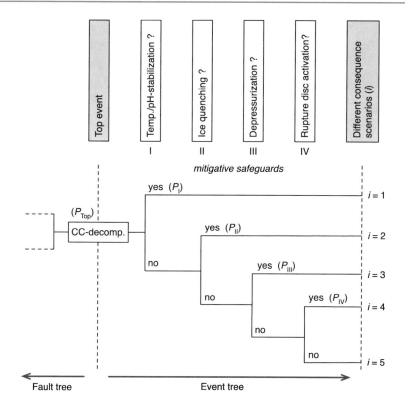

Fig. E.5 Event tree showing the pathway from the decomposition of cyanuric chloride to the resulting consequences considering mitigative safeguards. It is assumed that the reactor vessel is not able to handle the increase in pressure. P_{Top} = probability of top event; P_I to P_{IV} = reliability of mitigative safeguards; P_i = probability of consequence i

importance of the independence of safeguards (also regarding the electricity supply). Passive safeguards that do not rely on electricity can help ensure this independence.

Note that in fault trees, the variable P represents the probability of failure (fault) of system components, whereas in event trees, P represents the safety provided by these components. To help visually differentiate between the two, event trees use roman numerals for the subscripts of P.

Reaction Calorimetry

F

F.1   Basic Aspects of Calorimetry

As introduced in Chap. 8, *calorimetry* is an experimental technique that can be used to characterize the heat released or absorbed by a system from a chemical reaction. First established in the eighteenth century, it can be described using the simplified heat balance previously introduced in Sect. 8.2:

$$\dot{q}_{acc} = \dot{q}_r - \dot{q}_c \tag{F.1}$$

- $\dot{q}_{acc}$: rate of heat accumulation
- $\dot{q}_r$: rate of heat released by the reaction
- $\dot{q}_c$: rate of heat exchanged with the surroundings (cooling)

One of the fundamental aspects that distinguish the different calorimetric methods that exist is the mode applied for temperature control. Three of the common operating modes are:

- Adiabatic: the temperature increase results from the released heat of reaction—allows for the determination of the thermal runaway curve.
- Isothermal: the temperature of the sample is kept constant—allows for the determination of the conversion term of the reaction rate.
- Dynamic: the temperature of the sample is varied linearly over a specified temperature range—allows for the determination of the enthalpies of reaction.

In the case of an adiabatic calorimeter, the heat balance can be simplified as:

$$\dot{q}_{acc} = \dot{q}_r \qquad (\dot{q}_c = 0) \tag{F.2}$$

© The Author(s), under exclusive license to Springer Nature Switzerland AG 2021
K. Hungerbühler et al., *Chemical Products and Processes*,
https://doi.org/10.1007/978-3-030-62422-4

For calorimeters operating with isothermal and dynamic temperature control, the heat balance can be simplified to the case of an ideal heat flow:

$$\dot{q}_r = \dot{q}_c \qquad (\dot{q}_{acc} = 0) \tag{F.3}$$

The other fundamental aspect that characterizes different calorimetric methods is the scale of the sample. This is also an important factor in choosing an appropriate calorimetric method. For the investigation of decomposition reactions, *micro-calorimetry* is suitable since it can safely measure very small heat fluxes from a scaled-down version of the reaction mixture (e.g., in the milligram range).

In contrast, *macro-calorimetry* is a technique designed for samples in the gram range, and *bench-scale calorimetry* is for samples in the kilogram range. These allow for the running of experiments closer to industrial operating conditions and are more suitable for the study of desired reactions during process development and scale-up.

Comparing these techniques, micro-calorimetry is a more cost-effective and time-saving approach than the other two. It also presents advantages from a safety perspective given the small reaction sizes. However, during micro-calorimetry only the temperature can be controlled, meaning that the effect of other variables such as stirring, dosing, etc., cannot be assessed using this technique.

For an accurate study of the desired reaction under real process conditions, macro- or bench-scale calorimetry should be the method of choice.

In the next section, the basic principles of *differential scanning calorimetry* are introduced as a micro-calorimetry technique widely used in the field of process safety for screening purposes. Further general information about calorimetry and other specific calorimetric techniques can be found in Stoessel (2008).

F.2   Differential Scanning Calorimetry

Differential scanning calorimetry (DSC) is a micro-calorimetric technique used for the determination of heats of reaction (e.g., for both desired and decomposition reactions) and of the approximate temperature intervals in which these reactions take place. DSC is, therefore, a suitable technique for the identification of thermal hazards at the early stages of process development.

The sample size in DSC is usually between 1 and 20 mg, and the temperature control mode can be dynamic or isothermal. The differential calorimeter consists of a temperature-controlled oven containing a crucible holding a sample of the substance as well as a reference crucible, as shown in Fig. F.1.

During the measurement, the temperature difference (ΔT) between the two crucibles is recorded as a function of time, with the temperature of the heating element either remaining constant (isothermal mode) or increasing over time (dynamic mode). ΔT becomes proportional to the thermal response of the sample. Closed and pressure-resistant crucibles are used to avoid material and thermal distortions caused by the evaporation of volatile components.

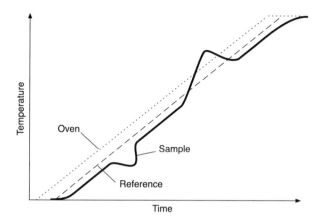

Fig. F.1 A generic schematic of results from a DSC showing the change in temperature over time for the sample, reference, and the oven (dynamic mode). Adapted from Gygax (1993)

F.2.1   Dynamic Mode

To measure the heat of reaction and the heat of decomposition, the reactants are mixed at low temperature and then heated linearly over time. This use of dynamic mode allows for the analysis of the whole temperature range of interest within a relatively short time. A scan over a temperature range of 20 °C (ambient temperature) up to 500 °C (at which most organic compounds decompose) can take approximately 2 hr.

The thermogram in Fig. F.2 shows the rate of heat release by the chemical reaction ($\dot{q}_r$) and by the decomposition reaction ($\dot{q}_d$) as a function of time or temperature. Integration of the rate of heat release over time yields the heat of the desired reaction (Q_r) and the decomposition reaction (Q_d).

The resulting enthalpies can be used to estimate the corresponding adiabatic temperature rise according to Eq. 8.10.

F.2.2   Isothermal Mode

The rate of heat release by the decomposition reaction ($\dot{q}_d$) at T_0—the maximum temperature of the synthesis reaction (MTSR)—is necessary for the characterization of the time to maximum rate under adiabatic conditions (TMR_{ad}). However, it cannot be read directly from a DSC thermogram.

Consequently, the TMR_{ad} can be determined indirectly through calculation using the Arrhenius law, which requires the empirical estimation of the reaction activation energy (E_A).

This activation energy can be determined by a series of DSC experiments operated in isothermal mode, where the heat of the decomposition reaction is measured at different temperatures. By performing a linear regression between the

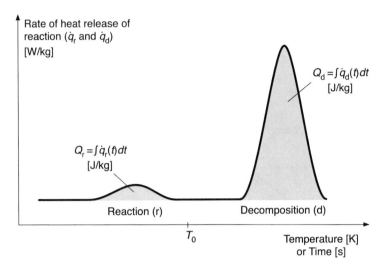

Fig. F.2 Example of a DSC thermogram of a reaction mixture including the decomposition. Note that due to thermal inertia, the onset points for the reaction and decomposition can be significant in determining the safety

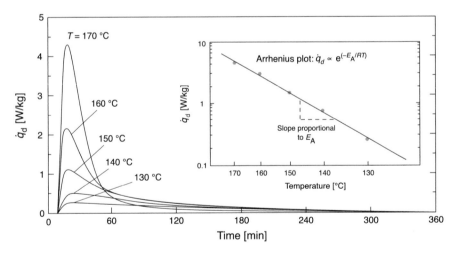

Fig. F.3 Example of an isothermal DSC thermogram with an Arrhenius plot of the thermokinetics of the decomposition reaction

natural logarithms of the maximum rates of reaction heat release and the reciprocal absolute temperatures, the activation energy can be obtained as the slope of the line. This methodology is illustrated in Fig. F.3.

Through these results, the rate of heat release of the decomposition reaction ($\dot{q}_d$) as a function of temperature can now be calculated using the activation energy (E_A)

and one reference point with the following expression:

$$\dot{q}_d(T_0) = \dot{q}_{ref} \times \exp\left\{\frac{-E_A}{R}\left(\frac{1}{T_0} - \frac{1}{T_{ref}}\right)\right\} \tag{F.4}$$

Then, the TMR_{ad} can be estimated through insertion into Eq. 8.12 as a function of temperature as follows:

$$TMR_{ad} = \frac{c_p}{\dot{q}_{ref} \times \exp\left\{\frac{-E_A}{R}\left(\frac{1}{T_0} - \frac{1}{T_{ref}}\right)\right\}} \times \frac{R \times T_0^2}{E_A} \tag{F.5}$$

By replacing TMR_{ad} with a value of 24 hr, the value of $T_0 = T_{24}$ can be determined. See Sect. 8.3.4 for additional information.

Case Study Solutions

G

Solutions for the tasks and questions within the case study (Chap. 10) are provided here. Some tasks and questions are rather qualitative and have no single correct answer. For qualitative tasks and questions, an example solution is provided but is not necessarily comprehensive.

G.1   Introduction

(a) Plant protection products are a subset of pesticides that work to prevent crops from being destroyed by disease or infestation. Pesticides can include herbicides, fungicides, insecticides, acaricides, plant growth regulators, and repellents (European Commission, 2019b; EFSA, 2019). Such products can help to prevent losses due to disease or infestation of crops.

(b) Plant protection products can be desirable to protect the health of a crop. Their active ingredients aim to work as intended on their target and through a specific mode of action. Undesired properties can include unintended side effects they might have on nontarget organisms including plants, animals, and humans in the surrounding environment. Products that are persistent beyond their intended application are mobile in multiple environmental compartments and/or degrade into substances that are mobile may also be problematic.

(c) Bion® aims to control fungal, bacterial, and viral plant diseases through a mode of action that mimics the natural systemic activated resistance (SAR) response found in most plant species (Syngenta, 2018). It works by boosting the plant's own resistance rather than by targeting the disease itself. This decreases the likelihood that the pathogen will develop increased resistance.

© The Author(s), under exclusive license to Springer Nature Switzerland AG 2021   303
K. Hungerbühler et al., *Chemical Products and Processes*,
https://doi.org/10.1007/978-3-030-62422-4

G.2   Process Risk Assessment

G.2.1   Step 2: Definition of Safe Process Conditions

(a) Yes, from the data collected in step 1, it is clear that HCN and methyl ethyl
    ketone are highly flammable, 3-methylpyridine is flammable, and 3-chloro-2-
    isopropylthio aniline reaches the flash point during the synthesis reaction. To
    ensure safe process conditions, ignition sources should be removed, and the
    reactor should be inerted.

    These safety measures should be applied during the entire process step (in
    both the synthesis and precipitation reactions).

(b) Under acidic conditions, CuCN and Na_2S can react to form HCN and H_2S,
    respectively. Both substances are highly toxic and flammable. In addition, HCN
    is formed during the precipitation reaction with Na_2S. Safe process conditions
    require that the presence of acidic conditions in the reactor is avoided and
    permanent monitoring of HCN levels.

(c) During the synthesis reaction, the temperature is controlled by the reflux
    conditions. During the precipitation reaction, the temperature is controlled by
    adjusting the addition rate of Na_2S.

    The synthesis reaction takes place at 190 °C due to a large activation barrier.
    Operating at 10 °C below the boiling point of the reaction mixture ensures
    process safety by avoiding the excessive formation of vapors, which might
    exceed the capacity of the condenser leading to pressure build-up and difficulties
    to control the temperature. Another aspect to keep in mind is that operating
    below the boiling point is also more energy efficient.

(d) The results of the thermal risk analysis are presented in Table G.1. From the data
    provided in step 1 in Table 10.1, it can be seen that 3-chloro-2-isopropylthio
    aniline and 3-amino-2-isopropylthio benzonitrile do not decompose when the
    MTSR of the synthesis reaction is reached.

Table G.1 Results of the thermal risk analysis for the synthesis reaction and work-up of 3-amino-
2-isopropylthio benzonitrile. The impact classification and criticality class are defined in Table 8.3
and Fig. 8.5, respectively, in Chap. 8

	Synthesis reaction	Precipitation reaction
Adiabatic temperature rise [°C]	95	55
MTSR [°C]	285	120
Impact(s) in case of adiabatic temperature rise	Boiling point reached and pressure build-up	Boiling point reached and pressure build-up
Impact classification	Medium	Medium
Criticality class	3	3

G.2.2 Step 3: Risk Identification

Table G.2 shows the completed HAZOP study table.

G.2.3 Step 4: Risk Analysis

Consequence Analysis

(a)

$$c_{max}(x) = \frac{\dot{M}}{\pi \times u \times \sigma_y \times \sigma_z} = \frac{5.6 \times 10^{-4}\,\frac{kg}{s}}{\pi \times 1.5\,\frac{m}{s} \times 18\,m \times 9\,m} = 7.3 \times 10^{-7}\,\frac{kg}{m^3}$$

Multiplying this concentration by one million and dividing by the average density of air leads to an estimation of the HCN concentration in units of parts per million:

$$\frac{7.3 \times 10^{-7}\,kg\,HCN/m^3}{1.3\,kg\,air/m^3} \times 1{,}000{,}000 = 0.6\,ppm$$

Considering a short-term exposure limit for HCN of 3.8 ppm (ECHA, 2019b) and that the estimated concentration at the neighboring site derived from the turbulent Gaussian model is 0.6 ppm, the effect on people at the neighboring site can be classified as negligible or potentially as low (minor injury).

(b) Considering the toxicity hazards of HCN and H_2S, the effects of an HCN release on operators in the plant are classified as high (injuries with irreversible damages).

Probability Analysis

(a) Figure G.1 shows a complete version of the fault tree analysis.
(b) The assumption that the events in the fault tree are independent is not correct since a power failure could cause all electrical equipment to fail. This would lead to an increase in the probability of HCN release. The events that would be triggered by a power failure include alarm failure, active safeguard failure, reflux failure, stirrer failure, etc. In this case, a more thorough analysis should be carried out using, e.g., a common cause failure analysis (see Appendix E.3).
(c) Table G.3 shows the results of a qualitative probability analysis for the release of HCN.

Table G.2 Results of the HAZOP study of the synthesis reaction and work up of 3-amino-2-isopropylthio benzonitrile. This Table does not represent an exhaustive list

Item	Process conditions	Deviation	Process hazard	Process failure	Loss event	Safeguards	Consequences
Synthesis reaction	Process temperature = 190 °C	High temperature	1. Fire and explosion hazards: HCN, 3-methylpyridine, 3-chloro-2-isopropylthio aniline[a] 2. Toxicity hazards: HCN is highly toxic, 3-methylpyridine and 3-chloro-2-isopropylthio aniline can be classified as dangerous substances 3. High process temperature 4. Thermal hazards: boiling point of reaction mixture is reached in the case of adiabatic temperature rise	1. Cooling failure: reflux failure, stirrer failure 2. Too much heating: thermometer failure, operator failure 3. Concentration too high: overcharge of reactants, undercharge of solvents 4. Power failure	Thermal runaway	1. Temperature alarm (preventive) 2. Depressurization/rupture disk activation (mitigative)	HCN release
Synthesis reaction	Atmospheric pressure	High pressure	1. Fire and explosion hazards: HCN, 3-methylpyridine, 3-chloro-2-isopropylthio aniline[a] 2. Toxicity hazards: HCN is highly toxic, 3-methylpyridine and 3-chloro-2-isopropylthio aniline can be classified as dangerous substances 3. High process temperature 4. Thermal hazards: boiling point of reaction mixture is reached in the case of adiabatic temperature rise	1. Cooling failure: reflux failure, stirrer failure 2. Too much heating: thermometer failure, operator failure 3. Concentration too high: overcharge of reactants, undercharge of solvents 4. Power failure	Rupture/crack hole in reactor	1. Temperature alarm (preventive) 2. Depressurization/rupture disk activation (mitigative)	HCN release
Precipitation reaction	Feed flow (Na$_2$S) approximately 105 kg/hr[b]; Process temperature = 65 °C	High feed flow (semi-batch); high temperature	1. Fire and explosion hazards: HCN, methyl ethyl ketone, H$_2$S 2. Toxicity hazards: HCN and H$_2$S are highly toxic, Na$_2$S and 3-amino-2-isopropylthio-benzonitrile can be classified as dangerous substances 3. Thermal hazards: at a process temperature higher than 25 °C, the boiling point of the reaction mixture would be reached in case of adiabatic temperature rise	1. Cooling failure: cooling system failure, stirrer failure 2. Feed flow (Na$_2$S) too high: flowmeter failure, operator failure 3. Power failure	Thermal runaway	1. Temperature alarm (preventive) 2. Depressurization/rupture disk activation (mitigative)	HCN release

[a] 3-Chloro-2-isopropylthio aniline reaches its flash point at reaction conditions
[b] Depending on the temperature control loop

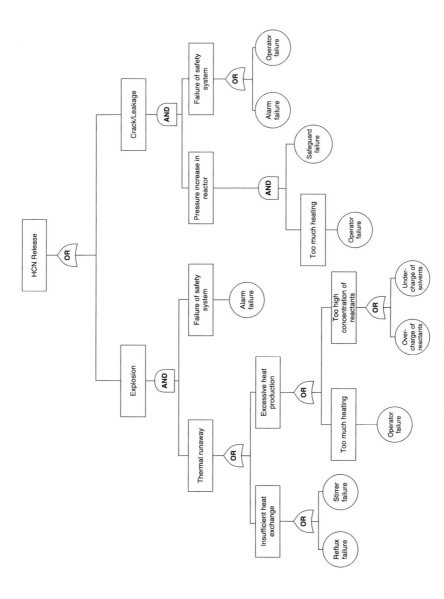

Fig. G.1 Result of the fault tree analysis concerning the release of HCN

Table G.3 Possible results for a qualitative probability analysis of the synthesis reaction and work-up of 3-amino-2-isopropylthio benzonitrile. The qualitative estimate of probability for each entry is shown in italics. Note that this does not represent an exhaustive list

Process failure	Process deviation	Safeguard failure (preventive)	Loss event	Safeguard failure (mitigative)	Consequence
• Failure of temperature measurement (*medium*) • Stirrer failure (*medium*) • Reflux failure (*medium*) • Operator failure (*medium*) • Flowmeter failure (*medium*) • Failure in the electricity supply (*medium*)	• Temperature too high (*medium*) • Pressure too high (*medium*)	• Failure of temperature alarm (*medium*)	• Thermal runaway (*medium*)	• Depressurization/ rupture disk activation (*medium*)	• HCN release (*medium*)

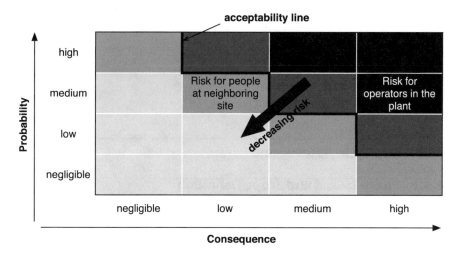

Fig. G.2 Risk matrix for the release of HCN in the work-up reaction of 3-amino-2-isopropylthio benzonitrile

G.2.4   Step 5: Risk Evaluation

As determined in the previous section and shown in Fig. G.2, the risk for people at the neighboring site is acceptable (below the acceptability line). However for operators at the plant, the risk level is not acceptable, and additional safety measures need to be implemented in order to reduce the risk to an acceptable level.

G.2.5   Step 6: Risk Management

(a)

- Make use of an emergency power aggregate for all critical process equipment and an emergency cooling unit.
- Different safety mechanisms should not be coupled to the same power source, but instead have independent emergency power sources.
- Ensure periodic maintenance of the reactor, measurement devices, and alarms. Have an inspection and maintenance plan.
- Install permanent monitoring of workplace air for HCN and H_2S levels with a connected alarm.
- Have personal protection equipment ready in case of an accidental release and for any sampling exercise where exposure could occur.
- Create an emergency plan and complete regular safety training with operators.

- In addition, the installation of a depressurization valve and a rupture disk followed by an alkaline absorption column for the case of HCN and H_2S release should be evaluated.

(b)

- HCN is still produced as a by-product.

G.3   Life Cycle Assessment

G.3.1   Phase 1: Goal and Scope Definition

(a) The goal of the LCA is to identify the sources of the largest impacts considering climate change, toxicity, and cumulative energy demand for the current reaction design to produce 3-amino-2-isopropylthio benzonitrile.
(b) Based on the description provided, the system boundaries can be defined as illustrated in Fig. 10.3 and used to define the boundary for the process risk assessment. The reactants (including solvents), heating energy, and waste solvents are within the scope of this LCA.
(c) The production of one batch (290 kg) of the product 3-amino-2-isopropylthio benzonitrile can be used as the functional unit.
(d) The target audience of this LCA is internal members of the company, which likely includes process engineers and company management. While the process engineers familiar with the plant may have highly relevant technical knowledge in the field, management may not. Thus it is important to deliver the results of the LCA in a format that would be understandable and useful for both groups.

G.3.2   Phase 2: Inventory Analysis

Table G.4 shows the LCI table documenting the material and energy flows that move into and out of the set system boundary.

Table G.4 Mass and energy flows within the the system boundary to define the life cycle inventory

Material or energy flow	Amount
3-Chloro-2-isopropylthio aniline	303 kg
Copper(I) cyanide (CuCN)	200 kg
Sodium sulfide (Na_2S)	150 kg
3-Methylpyridine (clean)	210 kg
Methyl ethyl ketone (clean)	180 kg
3-Methylpyridine (waste)	210 kg
Methyl ethyl ketone (waste)	180 kg
Steam	224 MJ
3-Amino-2-isopropylthio benzonitrile	290 kg

G.3.3 Phase 3: Impact Assessment

Table G.5 shows the calculated impacts for each of the three indicators across all of the material and energy flows identified in the LCI. Remember that the characterization factors assumed to represent some of the material and energy flows may not accurately reflect reality, and this will be considered further in the final phase of interpreting the results.

G.3.4 Phase 4: Interpretation

(a) Table G.5 shows the total midpoint impacts for climate change, CED, ecotoxicity, and human toxicity for this process step assuming the waste solvents are incinerated. The production of 3-chloro-2-isopropylthio aniline and 3-methylpyridine have the highest greenhouse gas emissions and cumulative energy demand, whereas production of copper(I) cyanide is overwhelmingly responsible for the human toxicity and ecotoxicity.

(b) According to the treatment processes applied that incinerate the waste solvents, a negative impact occurs. This is due to the energy produced by the incineration offsetting the need for production of energy from other fuels. Considering the significant impacts generated from virgin production of the solvents, it is almost certainly worthwhile to more closely consider recycling the solvents rather than using them just once and then incinerating them. The credits received from incineration do not nearly offset the impacts from virgin production of these solvents.

(c) This LCA could be seen as a first, rough calculation to identify hotspots within this process step. Some of the limitations and uncertainties include:

- Due to data limitations, the characterization factors applied are not completely representative of the material flows being assessed. More specific data able to describe the specific reactants and solvents used in this case study are needed.
- More data are needed to describe the efficiency of recycling systems that could be applied for solvent reuse. Characterization factors would then be needed to model these impacts and assess options.
- Energy use for other equipment involved in the process has not been considered. Other energy sources could also be considered for heating the reactor. For example, waste solvents could potentially be incinerated on-site to support steam generation, or use of renewable energy sources could be implemented.
- The quality of the available LCI data could also be checked to ensure accuracy and completeness before making any investment decisions.
- The other waste streams produced in this process step (e.g., Cu_2S, NaCl, HCN) should also be considered in the LCA.

Table G.5 Total calculated impacts for each material and energy flow for each of the midpoint impacts. Total impacts for the production of all reactants, wastes, and heat used are also shown. Results are rounded to three significant figures

Material or energy flow	Category	Climate change IPCC 2013 100 yr [kg CO$_2$-eq]	Total CED [MJ-eq]	USEtox ecotoxicity [CTU$_e$]^a	USEtox total human toxicity [CTU$_h$]^b
3-Chloro-2-isopropylthio aniline	Reactant	1460	31,500	5670	5.03×10^{-4}
Copper(I) cyanide (CuCN)	Reactant	594	10,000	234,000	1.91×10^{-2}
Sodium sulfide (Na$_2$S)	Reactant	458	6350	2540	2.58×10^{-4}
3-Methylpyridine (clean)	Solvent	1930	39,100	34,900	9.28×10^{-4}
Methyl ethyl ketone (clean)	Solvent	326	11,200	648	5.58×10^{-5}
3-Methylpyridine (waste)	Waste	−426	−620	−183	-1.70×10^{-5}
Methyl ethyl ketone (waste)	Waste	−365	−531	−157	-1.46×10^{-5}
Steam	Heat	23.1	349	17.3	1.85×10^{-6}
Total for reactants and solvents		*4768*	*98,150*	*277,758*	*2.08×10^{-2}*
Total for wastes		*−791*	*−1151*	*−340*	*-3.16×10^{-5}*
Grand total (reactants, solvents, waste, and heat)		*4000*	*97,348*	*277,435*	*2.08×10^{-2}*

^{a} CTU$_e$ = comparative toxic units for ecotoxicity
^{b} CTU$_h$ = comparative toxic units for human toxicity

(d) From these results, it can be seen that the production of copper(I) cyanide has the most significant impact in regard to human toxicity and ecotoxicity. Efforts should be made to maximize the efficient use of this substance. Incinerating the waste solvents appears to not be a most efficient treatment method, and more data and modeling is needed to understand options for recycling and reuse.

G.4 Product Risk Assessment

G.4.1 Step 1: Hazard Identification

Yes, according to ECHA's brief profile (ECHA, 2019a), ASM is classified under the GHS as harmful (GHS07) and an environmental hazard (GHS09). Specifically, it is described as being very toxic to aquatic life with long-lasting effects, causing serious eye irritation, skin irritation, and possibly skin sensitization and may cause respiratory irritation.

G.4.2 Step 2: Exposure Analysis

Level I Environmental Fate Model

(a) By writing out the relationship of the two partition coefficients to the concentrations:

$$K_{sw} = c_s/c_w$$
$$K_{aw} = c_a/c_w$$

The concentrations in each of the three compartments can then be determined as:

$$c_s = m_s/V_s$$
$$c_w = c_s/K_{sw}$$
$$c_a = K_{aw} \times c_w$$

(b) Assuming a (1) total surface area of 10.18 trillion m^2, (2) fraction of area covered by water of 0.048 (4.8%), and (3) fraction of area covered by soil of 0.952, the level I distribution within the small-world model shows that the majority of ASM (near 82%) will partition to water. Less than 18% will partition to the soil, and less than 1% will be in the air.

Level III Environmental Fate Model

(a) The three equations are:

$$\frac{dm_s}{dt} = E_s - m_s \times k_s + t_{as} \times m_a - t_{sa} \times m_s + t_{ws} \times m_w - t_{sw} \times m_s$$

$$\frac{dm_a}{dt} = E_a - m_a \times k_a + t_{sa} \times m_s - t_{as} \times m_a + t_{wa} \times m_w - t_{aw} \times m_a$$

$$\frac{dm_w}{dt} = E_w - m_w \times k_w + t_{sw} \times m_s - t_{ws} \times m_w + t_{aw} \times m_a - t_{wa} \times m_w$$

(b) Using the information provided on ASM in Table 10.7, the continuous emission rate of ASM into 1 ha of soil during the growing season can be calculated as the total amount of ASM applied over the entire season of 120 days (assuming five applications with 24 days between each application):

$$\frac{75 \text{ g Bion}}{\text{ha}} \times \frac{500 \text{ g ASM}}{1000 \text{ g Bion}} \times \frac{5 \text{ applications}}{\text{season}} \times \frac{1 \text{ mol ASM}}{210 \text{ g ASM}} \times \frac{1}{2880 \text{ hr}}$$

$$= \frac{3.1 \times 10^{-4} \text{ mol ASM}}{\text{hr}}$$

(c) The resulting distributions for the level III solution from the small-region model using a continuous emission rate of 3.1×10^{-4} mol/hr are:

- For emission only to air: 54% air, 2% water, 44% soil, 0% sediment
- For emission only to water: 0% air, >99% water, 0% soil, <1% sediment
- For emission only to soil: 0% air, <1% water, >99% soil, 0% sediment
- For emission only to sediment: 0% air, <1% water, 0% soil, >99% sediment

(d) With the exception of the air compartment, the largest fraction (>99%) of ASM remains in the compartment where it was emitted. This shows the low propensity of ASM to be redistributed to other compartments after emission into the environment. Of the four compartments, ASM is expected to have the most mobility when emitted to air, particularly into the soil compartment.

(e) The level III model assumes a steady-state emission rate and is not able to predict changes in the compartment inventory over time in response to changing emission rates. It also does not consider other environmental compartments that exist such as vegetation, the various types of soil, etc.

The assumption of steady-state emission also does not most accurately represent the actual use of Bion® (and other crop protection products). The corresponding local environmental concentrations are better modeled with more complex, time-resolved models that give realistic peak concentrations in representative soils or water bodies.

Human Exposure: Occupational

(a) EFSA set the AOEL for ASM as 0.03 mg/kg body weight per day (EFSA, 2014). The maximum application rate of ASM is calculated to be 0.0375 kg/ha. The estimated acute and longer-term systemic exposures are summarized in Table G.6.

Table G.6 Estimated total systemic acute and longer-term occupational exposure to ASM according to the EFSA tool and applied scenarios [mg kg_{bw}^{-1} d^{-1}] considering the use of personal protection equipment (PPE)

Total acute systemic exposure		Total longer-term systemic exposure	
Without PPE	With PPE	Without PPE	With PPE
0.26	0.11	0.15	0.015

(b) Longer-term systemic exposure primarily results from exposure to the body during application.
(c) Wearing workwear that covers the arms and legs reduces the dermal (and total) exposure significantly. In comparison, the use of gloves and a hood and visor only slightly decreases exposure according to the model. Note that the Bion® product label requires this type of workwear to be worn by operators.

Human Exposure: Dietary

(a) Tables G.7 and G.8 show the average dietary intake values [g kg_{bw}^{-1} d^{-1}] for Italy and and the United Kingdom.

Table G.7 Dietary intake values [g kg_{bw}^{-1} d^{-1}] of various food categories in Italy for the three dietary groups of infant, toddler, and adult as published in Anastassiadou et al. (2019)

	Infant	Toddler	Adult
Apples	–	0.89	0.79
Apricots	–	0.10	0.11
Bananas	–	0.54	0.20
Hazelnuts	–	0.00	0.00
Lettuce (leaf vegetables)	–	0.61	0.81
Mangoes	–	–	–
Peaches	–	0.35	0.38
Pears	–	0.35	0.23
Spinach and similar	–	0.17	0.22
Tomatoes	–	1.43	1.16

(b) Table G.9 shows the set maximum residue levels (MRLs) in the EU for each food category.
(c) The calculated values for the predicted daily intake are shown in Table G.10.

Table G.8 Dietary intake values [$g\,kg_{bw}^{-1}\,d^{-1}$] of various food categories in the United Kingdom for the three dietary groups of infant, toddler, and adult as published in Anastassiadou et al. (2019)

	Infant	Toddler	Adult
Apples	1.56	1.71	0.41
Apricots	0.13	0.02	0.01
Bananas	1.46	1.08	0.35
Hazelnuts	–	0.01	0.00
Lettuce (leaf vegetables)	0.01	0.05	0.14
Mangoes	–	0.01	0.01
Peaches	0.05	0.07	0.03
Pears	0.25	0.18	0.06
Spinach and similar	0.01	0.03	0.02
Tomatoes	0.37	0.59	0.44

Table G.9 Set maximum residue levels [mg/kg] of ASM in the EU for various food categories as published in European Commission (2019a)

Food category	Set MRL
Apples	0.3
Apricots	0.2
Bananas	0.08
Hazelnuts	0.1
Lettuce (leaf vegetables)	0.4
Mangoes	0.6
Peaches	0.2
Pears	0.2
Spinach and similar	0.6
Tomatoes	0.3

(d) Yes, the intake does differ significantly between the foods (in some cases by a factor of 10 or more). In Italy, both toddlers and adults are estimated to consume the highest amounts of ASM from tomatoes, lettuce, and apples. In the United Kingdom, the highest consumption for all three age groups is estimated to come from apples. In both countries, toddlers are predicted to consume the highest total amount from this set of foods.

G.4.3   Step 3: Effect Assessment

Environmental Toxicity

Toxicity testing of ASM resulted in:

- An E_bC_{50} value of 500 µg/L during an acute 72-hr static test on the algae species *Scenedesmus subspicatus*
- a NOEC value of 44 µg/L during a 22-day semi-static test on the daphnia species *Daphnia magna*
- a NOEC value of 26 µg/L during an 87-day flow through test on the fish species *Oncorhynchus mykiss*

See page 58 in EFSA (2014).

Table G.10 Calculated dietary predicted daily intake (PDI) [$\mu g \, kg_{bw}^{-1} \, d^{-1}$] of ASM for a set of foods and age groups. Based on dietary intake data from the EFSA PRIMo model and set maximum residue levels (MRLs) for each food in the EU

	Italy			United Kingdom		
Food	Infant	Toddler	Adult	Infant	Toddler	Adult
Apples	–	0.27	0.24	0.47	0.51	0.12
Apricots	–	0.02	0.02	0.03	0.00	0.00
Bananas	–	0.04	0.02	0.12	0.09	0.03
Hazelnuts	–	0.00	0.00	–	0.00	0.00
Lettuce (leaf vegetables)	–	0.24	0.32	0.00	0.02	0.06
Mangoes	–	–	–	–	0.01	0.01
Peaches	–	0.07	0.08	0.01	0.01	0.01
Pears	–	0.07	0.05	0.05	0.04	0.01
Spinach and similar	–	0.10	0.13	0.01	0.02	0.01
Tomatoes	–	0.43	0.35	0.11	0.18	0.13
Total	–	1.25	1.20	0.79	0.88	0.38

Human Toxicity: Occupational

The published AOEL within the EFSA conclusion (EFSA, 2014) is $0.03 \, mg \, kg_{bw}^{-1} \, d^{-1}$.

Human Toxicity: Dietary

The published ADI within the EFSA conclusion (EFSA, 2014) is $0.03 \, mg \, kg_{bw}^{-1} \, d^{-1}$.

G.4.4 Step 4: Risk Characterization and Classification

Environmental Toxicity

Based on these EC_{50} and NOEC values for each species with an assessment factor of 10 and the steady-state PEC of ASM in water estimated by the small-region model ($8.9 \times 10^{-7} \, mol/m^3$ or $0.19 \, \mu g/L$), the risk quotient (RQ) can be calculated as follows:

$$RQ_{algae} = \frac{0.19 \, \mu g/L}{500 \, \mu g/L \times 1/10} = 0.0038$$

$$RQ_{daphnia} = \frac{0.19 \, \mu g/L}{44 \, \mu g/L \times 1/10} = 0.043$$

$$RQ_{fish} = \frac{0.19 \, \mu g/L}{26 \, \mu g/L \times 1/10} = 0.073$$

As the risk quotients are much less than one, there is likely not a risk posed to these species as indicated by the modeled environmental concentration value in this particular scenario and considering just these three toxicological studies. This result clearly represents a very simplified calculation based on limited data and cannot be used to assess the actual environmental safety of the product.

Human Toxicity: Occupational

Considering the set AOEL of $0.03 \, \text{mg} \, \text{kg}_{bw}^{-1} \, \text{d}^{-1}$ according to EFSA (EFSA, 2014), the calculated occupational exposure to operators without PPE is greater than 500% of the AOEL, and with PPE it is 52% of the AOEL. There is clearly a risk present when PPE is not used, and no risk is expected to be posed when PPE is worn.

EFSA writes that they arrived at this AOEL value (and also at the ADI value) "based on the LOAEL of $10 \, \text{mg/kg}_{bw}$ per day from the developmental toxicity study in rat, applying an uncertainty factor (UF) of 300 [referred to in this book as an assessment factor], 100 standard factor plus additional factor of 3 to account for the LOAEL basis" (EFSA, 2014).

Human Toxicity: Dietary

The highest calculated PDI for any population group and country (see Table G.10) was calculated as $0.00125 \, \text{mg} \, \text{kg}_{bw}^{-1} \, \text{d}^{-1}$. This is a factor of 24 below the set ADI of $0.03 \, \text{mg/kg}$ (EFSA, 2014), indicating a large margin of safety for dietary consumer risk.

G.4.5   Step 5: Risk Management

(a) Of the three areas investigated, the only risk that was found to be not acceptable was the occupational exposure to ASM for workers without PPE.
(b) PPE should always be properly worn when spraying Bion®.
(c) This case study was clearly limited in scope and made use of just a very small subset of data and available modeling tools to examine specific risks. Some of the clear limitations and uncertainties for each of the investigated areas include:

- Risk to the environment: (1) The small-world model is a simplified level III model making use of just a small set of physicochemical properties to describe ASM and the environment it could be applied within. Further, detailed information regarding the application and actual environmental conditions could be used in a more sophisticated model to estimate the PEC. Field studies and environmental monitoring would, of course, be more accurate. (2) Only a single toxicological study for just one aquatic species was considered in defining the PNEC. Many more toxicological studies, as well as a rigorous approach to applying a best assessment factor, are needed in order to define a more reliable PNEC.

- Occupational risk: (1) The report detailing the development and use of the EFSA tool (EFSA, 2014) should be reviewed to fully understand the assumptions made by the model used. (2) Assumed input data such as absorption levels for ASM may be more accurately refined through a review of recent information.
- Dietary risk: (1) Just a small subset of selected foods was investigated. Total dietary exposure should take into account all residues that could exist on foods within the diet. From the calculated margin of safety, a 24-fold higher dietary exposure would be required before the ADI is exceeded. (2) The calculations assumed that the maximum amount of the MRL was present on all samples (a worst-case scenario).

References

Atherton J, Gil F (2008) Incidents That Define Process Safety. John Wiley & Sons, Inc., Hoboken, NJ, USA, http://doi.wiley.com/10.1002/9780470925171

Anastassiadou M, Brancato A, Carrasco Cabrera L, Ferreira L, Greco L, Jarrah S, Kazocina A, Leuschner R, Magrans JO, Miron I, Pedersen R, Raczyk M, Reich H, Ruocco S, Sacchi A, Santos M, Stanek A, Tarazona J, Theobald A, Verani A (2019) Pesticide Residue Intake Model-EFSA PRIMo revision 3.1. EFSA Supporting Publications 16(3):1–15, https://doi.org/10.2903/sp.efsa.2019.EN-1605, http://doi.wiley.com/10.2903/sp.efsa.2019.EN-1605

CCPS (1999) Guidelines for Chemical Process Quantitative Risk Analysis, 2nd edn. John Wiley & Sons, Inc., https://www.wiley.com/en-us/Guidelines+for+Chemical+Process+Quantitative+Risk+Analysis%2C+2nd+Edition-p-9780816907205

CPD (1996) Methods for the calculation of Physical Effects. Tech. rep., http://content.publicatiereeksgevaarlijkestoffen.nl/documents/PGS2/PGS2-1997-v0.1-physical-effects.pdf

ECHA (2019a) Brief Profile: S-methyl benzo(1.2.3)thiadiazole-7-carbothioate. https://echa.europa.eu/brief-profile/-/briefprofile/100.101.876

ECHA (2019b) Hydrogen cyanide: Guidance of Safe Use. https://echa.europa.eu/registration-dossier/-/registered-dossier/14996/9

EFSA (2014) Conclusion on the peer review of the pesticide risk assessment of the active substance acibenzolar-S-methyl. EFSA Journal 12(8):3691, http://doi.wiley.com/10.2903/j.efsa.2014.3691

EFSA (2019) Pesticides. https://www.efsa.europa.eu/en/topics/topic/pesticides

Elvers B (ed) (2000) Ullmann's Encyclopedia of Industrial Chemistry. Wiley, Weinheim, Germany, http://doi.wiley.com/10.1002/14356007

ESCIS (1994) Inerting: Methods and Measures for the Avoidance of Ignitable Substance-Air Mixtures in Chemical Production Equipment and Plants. ESCIS Safety Series 3, https://doi.org/10.3929/ETHZ-B-000354483

European Commission (2019a) EU Pesticides Database. http://ec.europa.eu/food/plant/pesticides/eu-pesticides-database/public/?event=homepage&language=EN

European Commission (2019b) Pesticides. https://ec.europa.eu/food/plant/pesticides_en

Gygax R (1993) Thermal Process Safety. ESCIS Safety Series 8, https://doi.org/10.3929/ETHZ-B-000354503

IEC (2017) IEC 60079-0. http://dom1.iec.ch/publication/62232

Jacob DJ (1999) Introduction to Atmospheric Chemistry. Princeton University Press, https://press.princeton.edu/books/hardcover/9780691001852/introduction-to-atmospheric-chemistry http://acmg.seas.harvard.edu/people/faculty/djj/book/

Stoessel F (2008) Thermal Safety of Chemical Processes. Wiley-VCH Verlag GmbH & Co. KGaA, Weinheim, Germany, http://doi.wiley.com/10.1002/9783527621606

Syngenta (2018) Bion 500 FS Seed Treatment. http://www.syngenta-us.com/labels/bion-500-fs
UN (2019) Globally Harmonized System of Classification and Labelling of Chemicals (GHS), eighth edn. https://www.unece.org/trans/danger/publi/ghs/ghs_rev08/08files_e.html
US CSB (2020a) BP America Refinery Explosion. https://www.csb.gov/bp-america-refinery-explosion/
US CSB (2020b) Completed Investigations. https://www.csb.gov/investigations/completed-investigations/

Glossary

Active substance The specific chemical within a product formulation that serves as the active component in achieving the targeted effect (e.g., the chemical in a plant protection product formulation that achieves the effect needed to protect the crop from the intended disease or pest).

Assessment factor (AF) A factor applied during the calculation of a no-effect concentration within product risk assessment to consider the uncertainty and variability within and between different species as well as the time limitation or uncertainty of the applied toxicity data.

Batch process A type of chemical production process that follows a chronological execution of process steps in a single reaction vessel and according to a standard operating procedure. The desired chemical is produced in a specified amount and within a set time frame.

Batch reactor A type of reactor where all reactants are fed into the vessel at once.

Bioaccumulation factor The ratio of a chemical's concentration in an aquatic organism and in the surrounding water under the condition that uptake of the chemical occurs via all possible exposure routes.

Bioconcentration factor The ratio of a chemical's concentration in an aquatic organism and in the surrounding water under the condition that uptake of the chemical occurs via gill respiration, but not with food.

Biomagnification factor The ratio of a chemical's concentration in an aquatic or nonaquatic organism and in the food (prey) consumed by this organism.

Characterization factor (CF) A value used within life cycle assessment and, more specifically, in life cycle impact assessment to convert from a material or energy flow into a specific midpoint or endpoint impact category.

K. Hungerbühler et al., *Chemical Products and Processes*,
https://doi.org/10.1007/978-3-030-62422-4

Chemical process A system used to produce one or more chemical products. Within the scope applied in this book, it includes (1) the chemical reaction where raw materials (reactants) are converted to the desired product as well as to by-products and waste; (2) physical separation and purification operations such as crystallization, filtration, and distillation used to obtain the desired product specification; and (3) handling, transfer, and storage of chemicals and waste involved.

Chemical product A chemical that has been manufactured or extracted from a natural source to provide a specific function. In the context of this book, the focus is placed on chemical products as single chemicals. However, chemical products are also present on the market as formulations or mixtures of multiple chemicals.

Continuous process A type of chemical production process where the material flow is continuously in motion (without interruption) as it advances toward becoming the end product. Mostly used for products that are produced in large amounts (bulk chemicals).

Cumulative energy demand (CED) The total amount of both direct and indirect uses of primary energy involved in the extraction, manufacturing, and disposal of all raw and auxiliary materials for a product or process including the involved technical equipment (proportionally).

Damage The impairment of the inherent qualities of a protected good.

Danger A situation or circumstance in which damage may occur.

Decision As applied within the EU, a decision is a binding act that must be followed by the countries or entities it specifically addresses.

Directive As applied within the EU, a directive sets a goal that all EU member states must achieve, but each member state can establish their own laws to achieve it.

Dose-response relationship The relationship between the level of exposure to a chemical (the dose) and the resulting effect (response) it elicits on an investigated endpoint. This can include specific effects on both humans and the environment. In particular, dose-response relationships may also be non-monotonic, as it is observed for some endocrine-disrupting chemicals (EDCs).

Eco-efficiency One of the three guiding principles of integrated development (in addition to inherent safety and social acceptability). It aims to proactively minimize resource use and environmental impacts per unit of service provided by a chemical product or process rather than relying on reactive approaches such as waste treatment by end-of-pipe technologies.

Endocrine disrupting chemical (EDC) According to the World Health Organization, this is an "exogenous substance or mixture that alters function(s) of the endocrine system and consequently causes adverse health effects in an intact organism, or its progeny, or (sub)populations."

Exposure Describes the presence of a chemical or another potentially hazardous agent at a protected good. Exposure is quantified in terms of its magnitude (concentration or dose of the chemical) and its duration.

External cost A cost that occurs when someone obtains a benefit at the expense of other individuals or the greater society.

Formulated product A chemical product that contains a mixture of ingredients that are both *active substances* and inactive substances, including so-called auxiliaries.

Functional unit Used within life cycle assessment as the reference utility or service to which input and output flows and the associated impacts of a product or process are all related.

Global warming potential (GWP) (1) A property of greenhouse gases that indicates the heat absorbed by a gas relative to the heat absorbed by the same mass of carbon dioxide. (2) The characterization factor used within life cycle impact assessment to calculate climate change impacts. Often defined in terms of carbon dioxide as the reference substance and expressed in units of kg carbon dioxide equivalents.

Hazard An inherent property of a substance or situation with the potential to cause harm.

Independent protection layer A specific type of safeguard designed and managed to perform independently of any initiating cause or other protection layers.

Inherent safety One of the three guiding principles of integrated development (in addition to eco-efficiency and social acceptability). It aims to proactively limit hazards within a chemical product or process rather than rely on reactive approaches such as risk control.

Integrated development An approach to creating chemical products and processes that considers risks related to safety and the protection of human health and the environment throughout the entire life cycle and from the very first development steps. It makes use of inputs from multiple perspectives and areas of expertise, and it encourages iterative improvements. It is based on the three guiding principles of eco-efficiency, inherent safety, and social acceptability.

Legislation Legally binding standards developed in the common interest of society. In particular, legislation for safety and protection of human health and the environment in regard to chemical products and processes involves the development of a system of rules for handling chemical technology.

Life cycle A series of stages through which a chemical product or process passes during its lifetime. For a chemical product, this can include resource extraction and provision of energy, production, use/consumption, as well as disposal, recycling, and/or treatment of the wastes generated during all of these stages. For a chemical process, this can include resource extraction and provision of energy, process operation (including any involved transportation and storage), as well as disposal, recycling, and/or treatment of the waste generated during all of these stages.

Life cycle assessment (LCA) A standardized method that can be used to identify, quantify, and interpret the impacts on the environment caused by the resource consumption and emissions related to a product and/or process. According to the standards set by ISO 14040, a life cycle assessment should comprise the steps of goal and scope definition, inventory analysis, impact assessment, and interpretation.

Maximum residue level (MRL) The highest legally tolerated concentration of a pesticide residue that is allowed to remain on food or animal feed when placed on the market.

Mixture toxicity The overall toxic effect caused by the simultaneous exposure to a mixture of multiple toxic substances. While each chemical individually has its own toxic effect, in a mixture chemicals can interact in different ways to produce either an additive, synergistic, or antagonistic overall effect.

Multilateral environmental agreement (MEA) A legally binding agreement set up between three or more countries with the aim of achieving an environmental goal.

Multimedia mass-balance model A type of model used to describe and estimate the behavior of chemicals in the environment (their so-called environmental fate) and their resulting concentrations in environmental compartments (e.g., air, water, soil, sediment). Multimedia mass-balance models are based on mass-balance equations (incoming vs. outgoing flows) for all environmental compartments considered.

Net present value (NPV) The total present monetary value of a product or process considering cash flows along its life cycle. It takes into account an adjustable discount rate representative of costs associated with future potential risks, inflation, and opportunity costs.

Occupational exposure limit (OEL) The maximum legally allowed external exposure of an employee to a substance at the workplace. Derived on the basis of information on the substance's hazardous properties.

Opinion As applied in the EU, an opinion is a nonbinding statement issued by EU government institutions and committees to clarify their viewpoint on an issue.

Persistence A measure of the time required for the biological and/or chemical degradation of a chemical in the environment.

Precautionary principle According to the 1992 Rio Declaration, the precautionary principle states, "where there are threats of serious or irreversible damage, lack of full scientific certainty shall not be used as a reason for postponing cost-effective measures to prevent environmental degradation."

Process risk assessment A method containing six steps (from scope and system definition to risk management) that can be used to identify, evaluate, and manage critical risks within chemical processes.

Product risk assessment A method containing five steps (from hazard identification to risk management) that can be used to identify, characterize, and manage the hazardous properties of and exposure to a chemical product.

Protected good Valued subjects, resources, or conditions such as humans, the environment, and property.

Protection layer A method within functional process safety used to define the order (or sequence) in which safeguards are implemented in the system.

Quantitative structure-activity relationship (QSAR) A quantitative relationship between a biological activity (such as toxicity) and a descriptor(s) of chemical properties used to predict the activity. QSAR models are regression models used to predict the biological activity of a chemical often through analysis of its structural features (e.g., functional groups, molecular shape). They can be used to help screen chemicals for potential toxicity.

REACH EU regulation on industrial chemicals; the acronym stands for the Registration, Evaluation, Authorization, and Restriction of Chemicals (Regulation (EC) No 1907/2007). The underlying principle of the regulation is to place the burden of generating data needed to prove a substance is safe on the producer or importer of the substance.

Recommendation As applied in the EU, a recommendation is a nonbinding statement that serves as a way for a government entity to make its views on an issue known.

Regulation As applied within the EU, a regulation is a binding act that has to be fully applied as it is written across all EU member states.

Reliability A quantity that describes the ability of a component or a system to work according to its specifications for a certain period of time. Specifically, the function $R(t)$ indicates the probability that the component or system is still working at time t.

Residual risk The risk(s) that remain even after completion of a risk assessment and implementation of risk management measures. This can include risks accepted considering costs and benefits, risks incorrectly assessed due to uncertainties, and risks that exist but have not been identified.

Risk In the context of chemical products and processes, a risk is the potential (probability) for harm (impact) to protected goods by specific, hazardous conditions, or circumstances.

Safeguard Within a chemical process, a safeguard is any engineered system or administrative control used to (1) interrupt the sequence of events following an initiating event (known as a *preventive* safeguard) or (2) mitigate the consequences of the events (known as a *mitigative* safeguard).

Scenario A set of hypothetical but possible situations or conditions that are used to describe a possible course of events. A scenario describes events that may occur in the future, but it does not constitute a prognosis.

Social acceptability One of the three guiding principles of integrated development (in addition to eco-efficiency and inherent safety). It aims to proactively ensure that a societal dialogue on the risks and benefits of a chemical product or process takes place rather than relying on reactive management of societal conflicts.

Substance of very high concern (SVHC) Under the European chemicals regulation REACH, an SVHC is a chemical with certain hazardous properties such as carcinogenicity, mutagenicity, or toxicity to reproduction (CMR properties) or persistence, bioaccumulation potential, and toxicity (PBT properties). All recognized SVHCs are published on the "Candidate List of substances of very high concern for Authorisation."

Sustainable development As defined in the 1987 Brundtland Report, it is "development that meets the needs of the present without compromising the ability of future generations to meet their own needs."

Thermal process safety A field within process risk assessment pertaining to the management of hazardous thermal energy that can be generated during industrial chemical reactions.

Uncertainty Lack of knowledge about the state of a system or the exact value of a property. Uncertainty is caused by many different factors that may contribute to the overall uncertainty surrounding the result of a scientific investigation. Uncertainty can be reduced by improved data availability and measurement methods. It is important to differentiate uncertainty and (natural) variability. Variability is a characteristic of a system and cannot be reduced by improved measurement methods. Variability is inherent to virtually all natural systems and their properties. Examples relevant in the context of this book include the susceptibility of organisms to a toxic chemical and the weather conditions affecting the environmental distribution of a chemical.

Index

Printed in the United States
by Baker & Taylor Publisher Services